Attachment, Climate Crisis and the Natural World

Attachment, Climate Crisis and the Natural World is the first book to consider the psychological and social impact of global warming through the lens of Attachment Theory. The contributors reflect on our relationship with the natural world from the perspective of Bowlby's major themes, including the nature and primacy of the attachment instinct for survival, the secure base, separation anxiety, maternal deprivation, loss and mourning. The therapeutic value of renewing our relationship with, and respect for, the natural world we are part of is explored, as is working with psychotherapy clients who are troubled by the climate emergency and suffering caused around the world by continued reckless plundering of natural resources. *Attachment, Climate Crisis and the Natural World* will offer valuable insights for psychoanalysts, psychotherapists, counsellors, psychologists, environmental activists, and anyone concerned about our relationship with the natural world in times of ecological crisis.

Linda Cundy is an attachment-based psychoanalytic psychotherapist, supervisor, independent trainer and author. She is the Attachment Theory Consultant to the Bowlby Centre in London.

The Bowlby Centre Monograph Series

A series of books taken from the Annual John Bowlby Memorial Conference, produced in association with The Bowlby Centre, London.

Titles in the series:

Shattered States
Disorganised Attachment and its Repair
Edited by Kate White

Trauma and Attachment
Edited by Sarah Benamer and Kate White

Unmasking Race, Culture, and Attachment in the Psychoanalytic Space
Edited by Kate White

Sexuality and Attachment in Clinical Practice
Edited by Kate White and Joseph Schwartz

Touch
Attachment and the Body
Edited by Kate White

Shame Matters
Attachment and Relational Perspectives for Psychotherapists
Edited by Orit Badouk Epstein

Dissociative Identities
Attachment-based Approaches to Psychotherapy
Edited by Sue Richardson, Melanie Goodwin, Emma Jack and Michele Jowett

Attachment, Climate Crisis and the Natural World
The John Bowlby Memorial Conference Monograph 2024
Edited by Linda Cundy

For further information about this series please visit https://www.routledge.com/The-Bowlby-Centre-Monograph-Series/book-series/KARNBCM

"This book provides hope for all of us working to embed environmental sustainability in our work. We are urged to put aside our unhelpful (albeit understandable) desire to blame others or to feel despondent. The authors, in lively prose, show how attachment theory can help us understand with compassion those in denial. And how we can heal our planet, with actions underpinned by healthier relationships with the natural world, others and ourselves."

Daniel Harwood, *consultant psychiatrist and Chair, Royal College of Psychiatrists Planetary Health and Sustainability Committee*

"This collection is a vital addition to the growing library of resources available to climate aware psychotherapists, adding attachment-based perspectives to the field of climate psychology. A range of contributors, including speakers at the 2024 Bowlby Memorial Conference, make crucial links between attachment styles and our loss of connection and belonging, disconnection from our roots in land and culture, and our inability to confront the existential crisis we face. The reflections honour Bowlby and his commitment to the economic, environmental and relational realities of the individual within society."

Tree Staunton, *climate psychologist and UKCP Hon Fellow*

"This remarkable collection brings together leading attachment-oriented psychotherapists to illuminate one of the most urgent clinical challenges of our time: helping people face climate-related fear, grief, and disorientation. Each chapter offers a distinct yet complementary lens, from Jeremy Holmes' ecological reframing of attachment to James Angel's bold questioning of psychotherapy's relevance in a burning world. Through explorations of nature, art, story, loss, and belonging, the authors provide grounded, compassionate guidance. This is an essential, wise, and hopeful resource for clinicians, with a masterful postscript by Linda Cundy."

Howard Steele, *PhD, Professor of Psychology, The New School; Co-Director*

Attachment, Climate Crisis and the Natural World

The John Bowlby Memorial Conference Monograph 2024

Edited by Linda Cundy

Routledge
Taylor & Francis Group
LONDON AND NEW YORK

Cover image: *from* Pendle Witches, Flood, 1996, Paula Rego. Tate, Purchased 1997. © Paula Rego. Photo: Tate

First published 2027
by Routledge
4 Park Square, Milton Park, Abingdon, Oxon OX14 4RN

and by Routledge
605 Third Avenue, New York, NY 10158

Routledge is an imprint of the Taylor & Francis Group, an informa business

British Library Cataloguing-in-Publication Data
A catalogue record for this book is available from the British Library

ISBN: 978-1-041-22785-4 (hbk)
ISBN: 978-1-041-22784-7 (pbk)
ISBN: 978-1-003-73394-2 (ebk)

DOI: 10.4324/9781003733942

Typeset in Times New Roman
by SPi Technologies India Pvt Ltd (Straive)

Contents

Acknowledgements

In 2023, when I suggested the theme for the following year's Annual John Bowlby Memorial Conference, Elizabeth Balgobin was CEO of the Bowlby Centre. She supported the proposal eagerly and liaised with me for many months as I curated the event. Elizabeth stepped down from her role in January 2024 but planned to attend the conference. However, she became ill soon after handing over her role and, shockingly, she passed away on 8 March – just six weeks before it took place.

Her replacement, Jo Matthews, had to hit the ground running but found time to connect with me as last-minute details were organised, and I thank her for her enthusiasm for this project. Also, I want to thank Poppy Weston, Events Co-Ordinator at the Bowlby Centre, for her input.

Of enormous value as I prepared the conference, and subsequently in putting together this book, has been the support of Kate White. Kate is Series Editor for all Bowlby Centre monographs and also holds the history of the organisation since its establishment (initially as the Institute for Self Analysis in 1986), having been involved from the start. I was able to turn to her for information about the founder, John Southgate. Not only did she confirm that his radical politics included strong and active environmental involvement, but she was also able to provide me with historical documents in corroboration. This validated my instinct that, as attachment-based psychoanalytic psychotherapists, our remit extends beyond the consulting room; it is our legacy to work toward creating a fairer, more secure society in whatever ways we can, and promoting the protection of our environment is an essential and urgent imperative.

Finally, thanks go to Susannah Frearson at Routledge for her input and support in the preparation of *Attachment, Climate Crisis and the Natural World*, the book.

Contributors

James Angel, PhD, UKCP, MBACP, is an attachment-based psychoanalytic psychotherapist and Bowlby Centre member. He is also a campaigner and researcher with a special interest in climate change and social justice.

Rebecca Attoe is a trainee counselling psychologist working at the South London and Maudsley NHS Foundation Trust, providing trauma-informed psychological care to individuals experiencing post-traumatic stress disorder (PTSD) resulting from adverse life events. She supports the delivery of the Grounding Project at Roots and Shoots alongside Myriam Sarens and brings her experience of trauma work, particularly with asylum seekers and refugees.

Linda Cundy is an attachment-based psychoanalytic psychotherapist and clinical supervisor in private practice in North London. She has 30 years of experience teaching in the field of counselling and psychotherapy, is an independent trainer and conference speaker, and has contributed to and edited five books to date, including *Attachment, Relationships and Food: From Cradle to Kitchen* (Routledge, 2021) and *Narratives of Attachment, Danger and Survival* (Karnac, 2026). Linda is the Attachment Consultant to the Bowlby Centre.

Roger Duncan, MSc, UKCP, is a registered systemic family therapist, systemic supervisor and author. He originally studied biology and has been involved in working with adolescents using nature-based practice for more than 30 years. He was one of the pioneer tutors of the Ruskin Mill Education Trust, where he was involved in the development of therapeutic education programmes for adolescents with complex and challenging behaviour in the woodlands and wilderness settings and had a leadership role in senior management. Roger has worked for more than ten years in CAMHS with adolescents who have experienced complex trauma, including unaccompanied asylum-seeking children (UASC) and also works in private practice, with individuals, families, and organisations. He is a visiting lecturer in Psychological Therapies for the Tavistock and Portman NHS Trust. Roger writes and lectures on the epistemology and practice of eco systemic psychotherapy and was the course director of the 2022 Confer Diploma Eco Psychotherapy and the Emerging Adolescent Mind. He is author of *Nature in Mind: Systemic Thinking and Imagination in Ecopsychology and Mental Health* (Routledge, 2018), and *Nature and Belonging: Navigating from Separation to Reattachment to the Great Mother* (Karnac, in press).

Jeremy Holmes, MD FRCPsych MRCP, is a retired medical psychotherapist, consultant psychiatrist and honorary Professor at the University of Exeter, where he set up the psychoanalytic studies programme in 2000. Recent books include *Attachment in Therapeutic Practice* (SAGE, with Arietta Slade, 2017) and *The Brain has a Mind of its Own: Attachment, Neurobiology and the New Science of Psychotherapy* (Confer Books, 2020). He now devotes as much time as possible to Green politics.

Catherine Jacobs, CPsychol, is a Chartered Research Psychologist and Associate Fellow of the British Psychological Society. She works on a freelance basis, holding an honorary position with the Centre for Abuse, Trauma and Suicide Studies (CATSs) at Middlesex University, London, with whom she has worked for many years. Her research and training expertise extend across social and developmental arenas, where she specialises in evidence-based interview assessments of attachment style, parenting and early experiences of care and abuse to assess and chart change and inform best practice in social care and academic settings both nationally and internationally. She has also recently been accredited with the EMCC as a Senior Practitioner Executive Coach, training with the Tavistock Institute in their renowned and in-depth systems-psychodynamic methodology.

For over two decades, Catherine has also had a parallel career as a visual artist, undertaking residencies and exhibitions.

Jane Ryan trained as a psychotherapist at the Centre for Attachment-based Psychoanalytic Psychotherapy (CAPP, now the Bowlby Centre) in the 1990s. She subsequently founded Confer and worked as its director from 1999 until her retirement in 2022. Confer was created to bridge gaps between different schools of thought in psychotherapy, offering a platform to the most innovative author-theorists in the field. Driven by Jane's own concerns over ecological crises, Confer worked closely with the ecopsychology community to ensure that our relationship with nature was given a much stronger voice in psychotherapeutic literature.

She is the editor of three books, *How Does Psychotherapy Work?* (Karnac, 2006), *Tales of Psychotherapy* (Karnac, 2007) and, with Roz Carroll, *What is Normal?* (Confer Books, 2020). Her nature essay "Field with a View" was published by Little Toller in 2025.

Jane lives in rural Suffolk and has most recently completed the MA in Creative Non-Fiction Writing at the University of East Anglia.

Myriam Sarens is a UKCP-registered psychotherapist specialising in Eye Movement Desensitisation and Reprocessing (EMDR) therapy. She works within the South London and Maudsley NHS Foundation Trust, providing trauma-informed psychological care to individuals experiencing post-traumatic stress disorder (PTSD) resulting from adverse life events. She is the lead clinician for the NHS Grounding Project, a community-based therapeutic initiative supporting refugees and asylum seekers who have experienced trauma. Delivered in collaboration with Roots and Shoots in London, the project integrates culturally sensitive practices with evidence-based interventions to address the complex psychological consequences of forced migration, displacement and trauma-related distress.

Maggie Turp is a semi-retired psychotherapist and supervisor in private practice, and an independent trainer offering clinically themed workshops to counselling and psychotherapy groups across the UK. She is also an author of two books published by Jessica Kingsley. More recently, she has contributed a chapter to *Holding the Hope: Reviving Psychological and Spiritual Agency in the Face of Climate Change* (PCCS Books, 2023). She is an active member of the Climate Psychology Alliance and an editor of its *Explorations* e-Journal.

Dawn Whittaker is an honorary counsellor with the Grounding Project. She gained a Postgraduate Diploma in Refugee Care from Essex University in 2020.

Foreword

Jane Ryan

The authors of this collection of essays have been tasked with a really complex assignment: to draw out the links between John Bowlby's Attachment Theory – originally developed to describe the qualities of the mother and baby relationship – and the psychological impacts of the ecological crisis on Earth. Where to begin? What might an understanding of relational bonds based on dependency tell us about humanity's wildly varying responses to climate change?

In 1969, the child psychiatrist John Bowlby began to postulate that the intimate subjective experience of the primary human relationship is central to our relational health throughout life. Since then, numerous attachment studies have confirmed this view. At the same time, since the 1980s when the impending dangers of the heating planet were becoming understood, people in related disciplines have been trying to make sense of attitudes towards the predicted environmental catastrophe. Theories abound in a wide range of cultural spaces, from sociology to discourse analysis, from sustainable energy to fiction, social psychology to activism. It makes complete sense that attachment theorists would join this ever more urgent enquiry. If we're going to solve or even just mitigate the climate crisis, we need joined-up thinking and collective cooperation – a unifying narrative. And, as several authors here have suggested, Bowlby would have approved. He was radical, ecological, and determined.

This book is, just as one would hope, an experimental search, an invitation to roam in other theoretical spaces – to examine how far Attachment Theory can be harnessed to gain understanding of the human relationship to the other-than-human – to creatures, plants and place. It is a quest to see if attachment behaviour can illuminate our gravest threat to survival: denial or disinterest in the ecological crisis and the failure to act; to see if it can help us to cope with whatever chaos lies ahead as we enter the Anthropocene, often now referred to as the Sixth Extinction.

As this book shows, attachment theorists have curiosity about how other disciplines are approaching this challenge. The psychotherapist-authors here are already influenced by environmental activism and green politics; already attuned to climate-related grief and anxiety (states of mind that are increasingly showing up in the psychotherapy consulting room but risk being dismissed as intrapsychic). Individually, they explore quite different cultural domains, searching for the fit between the relational and the fragile ecosystems upon which we depend. And, collectively, they demonstrate that any discussion about the psychology of climate change attitudes and attachment requires an expansive meta-view. This resulting book is – perhaps inevitably – a richly interdisciplinary text. The chapters flow between psychoanalysis, anthropology, Indigenous spirituality, fiction, economics, and political ideology. It's hard to imagine reading it without discovering something new. As one of the contributors Roger Duncan says, "When you start to look for things, when you start to look for them with intensity, interesting things turn up" (p. 30).

As Linda Cundy mentions in her preface, this project is an offshoot of a Bowlby Centre conference that took place in London in 2024. These gatherings have been running for over 25 years, offering a space for psychotherapists to present their insights on attachment-based psychotherapy, a modality that has been developing since 1988, when Bowlby offered supervision to the innovative psychotherapist John Southgate. Together, they developed a connection between the nuances of the caregiver-baby bond and a new approach to therapy that is centred on understanding the significance of that relationship. Out of those discussions, and with co-founder and collaborator Kate White, the Bowlby Centre emerged, offering a rigorous four-year training in attachment-informed therapy. The project has become an influential forum in the professional community for exploring and extending insight into the potency of attachment in all our intimate relationships. And, especially, the centrality of secure early bonds in our capacity to flourish.

The theoretical foundation of this enquiry rests on three aspects of Bowlby's findings: first, his observational analysis of human attachment styles, which he described as "secure," "anxious-avoidant," or "ambivalent." Second, the concept of the secure base and safety-seeking behaviour. Third, his understanding of the emotional force of separation and loss. Over time, new attachment categories have been added as researchers and psychotherapists have discovered more about how our first relational dependencies set up patterns for life. For example, "disorganised," which refers to the dysfunctional or "borderline" relational modes of those raised by a chaotic or abusive parent. "Dismissive," "fearful," and "preoccupied" relational styles have come to offer even more nuanced ways of understanding how failures in parenting establish certain relational patterns. Can these distinctive characteristics, which so deeply inform our emotional lives, also explain something about our responses to the ecological crisis? Is there a typical attachment profile for the climate change denier? Or the eco-activist? Can Bowlby's theory of interpersonal behaviour be mapped onto our relationship with the other-than human? Onto our attachment to the planet as a whole?

One of the assumptions underlying this project is that we don't just attach to each other but to the more-than-human as well – to animals, plants and place. We only have to think of the outcry at the axing of the 120-year-old sycamore poised on Hadrian's Wall, the sorrow of seeing orangutans stranded in decimated forests, or images of elephants dying of dehydration. Such sights evoke powerful emotions. We *feel* the suffering of other creatures. Studies led by the late behavioural psychologist Jaak Panksepp gave us an understanding that all mammals share a set of "primary affective systems": "SEEKING (expectancy), FEAR (anxiety), RAGE (anger), LUST (sexual excitement), CARE (nurturance), PANIC/GRIEF (sadness), and PLAY (social joy)" (Panksepp & Biven, 2012). Each of these innate systems is underpinned by specific neurotransmitters. For example, oxytocin in CARE, or dopamine in SEEKING. Most of us have the lived experience of feeling empathy for a helpless animal or love for a pet. We resonate across species.

The human ability to attach to other mammals with which we share certain nervous system structures is evident. But how might we attach to plants and to land? Perhaps most obviously because these are our primary source of sustenance and thus safety; they are the place we have evolved to know as home – an external secure base that is inflected with relational significance. The healing power of contact with the other-than-human is well documented and scientifically validated: time spent in green spaces or with other animals is known to improve mood and physical health, especially in reducing inflammation and blood pressure (see Weir, 2025). It supports recovery from the inflammatory disorders that are an increasingly serious problem within Western capitalist societies. Immersed in toxic diets, sedentary lifestyles, hyper-stimulation and stress, we clash with our ancient biology. Simply being Western is producing a plethora of illnesses and psychological distress. However, as some of the authors here show, nature-alienation can be countered by conscious engagement with the other-than-human, even in cities. It is wonderful that ecological contact is now available on prescription in the UK, described by the NHS as "green social prescribing."

Plants, of course, lack the nervous system that underpins human-to-human (or mammal-to-mammal) attachment, but they have been discovered to use electrical signals and organic compounds to respond to environmental stimuli, even to communicate distress or come to the assistance of each other (Miguel-Tomé & Llinás, 2021). They too are complex in their endeavour to exist, and central to our safety. Yet in industrialised societies, where clean water comes from a tap and food from supermarket shelves, we seem far from connected to the delicately balanced and highly sophisticated ecosystems that support us. We collectively act as though we can transcend ecology – have forgotten that without the finely tuned interactions between water, oxygen, light, soil, flora and fauna, we cannot live.

The healing potential of engagement with plants, soil and growing food is vividly shown in Linda Cundy's chapter on horticultural therapy, in which she considers what can be discovered from consciously connecting (or reconnecting) with Earth. She tells us about her visits to the *Grounding Project* in south London, where "biodiverse wildlife gardens offer a safe haven for both people and nature to flourish".[1] Described by UNESCO as "a place of peace and beauty for deprived multicultural communities," this project supports those recovering from severe trauma – often refugees from the Global South who have been torn from the original ecosystem in which they grew up. Here, Linda meets people who vividly recall the sensualities they have left behind – the sounds, smells, colours that made their homeland distinctive. Gardening at the Grounding Project aims to restore eco-sensibility. The participants put their hands back into the soil, and watch seeds grow into food. They experience a transitioning back into the body that is mediated by this contact with earth (and with Earth). Interestingly, when we touch healthy soil we are exposed to a bacterium called *Mycobacterium vaccae*. When this is taken into the body via the skin, it stimulates the production of the mood-enhancing and soothing hormone serotonin. Humans are, after all, biological creatures, a mass of interacting cells and our ancient biology thrives when we find ways to interact with the other-than-human. Crucially, though, the security offered by the Grounding Project is relational. As Jeremy Holmes writes in Chapter 1, "We have to think about the systemic field *between* individuals … Seen this way, the attachment model conceives ourselves as '*part of, and inseparable from nature*'" (p. 3).

As several of the authors (Duncan, Turp, Jacobs) suggest, Bowlby's formulations of attachment, separation and loss of the secure base might help us to find a position in relation to climate change as it transforms our world from the dependable into the chaotic. In Chapter 3, Roger Duncan, formerly a biologist and now a systemic ecopsychotherapist, offers a deeply personal story about his journey from experiencing "nature"[2] as a scientific matter to a profound shift of consciousness that arose from immersing himself in his local woods. "I built a relationship with the woodlands," he writes, "which was beyond merely identifying the plants. I knew the Latin name of the bluebells, *Hyacinthoides non scripta*, but I had never before had the experience of seeing bluebells grow and die away each spring for over ten years." He began to think of "nature" as a subject with agency that yields or pushes back, expands or retracts and with which one can form a relationship. Furthermore, Roger discovered that when he took young clients there they too began to heal from personal trauma. They were opening up to the "ecological unconscious" (Roszak, n.d.).

Perhaps this is why those engaged in ecological work, like Roger, are fascinated by the capacity of indigenous societies to live within the terms of their local ecosystem. Interestingly, the word "Indigenous" (from the *Latin indigena*) means "sprung from the land." These people have preserved a capacity for "deep systems thinking." Their lives are informed by awareness that what we do to the land or water has to be carefully considered; they exemplify a responsive and responsible attachment to place, honouring the other-than-human as subjective and sacred. Tragically, they face existential threat from climate-driven loss of habitable land, introduction of disease, assimilation or violence from those with vested interests in that territory: from modern colonialism. It is relevant to note here that the Global North is responsible for a staggering 92 per cent of excess CO_2 emissions, against which indigenous peoples are powerless (Pardikar, 2020). As Jeremy Holmes points out (p. 5), the Global North is exporting its climate problem to the South and to the places where climate change is hitting hardest. It's not insignificant that often these are also conflict-torn regions such as Syria, Chad and Sudan.

Clearly, in post-industrialised societies we have lost our implicit understanding that humanity depends on the maintenance of healthy ecosystems for life. We have embarked on an insecure relationship with Earth that is akin to the dismissive attachment style that underpins emotional independence. We know about the perils of climate change, yet act as though we are impervious to them. James Angel (chapter 2) suggests that we suffer from a profound dissociation that "points towards the fragmentation of the psyche. Parts of the self are split-off and disowned as 'not-me.'" (p. 18) The extraordinary lack of concern displayed across the Global North points to a collective psychopathology of denial.

Disavowal, which Maggie Turp considers even more dangerous, is the belief that "climate change is real but it's not a crisis; we still have time; technological solutions will be found." (p. 45) – that holding false hope that the ecological crisis will be solved distracts us from the urgency of the situation and from seeing where we can be most useful – in supporting ourselves and others in managing the effects as well as possible. (This, by the way, is exemplified by the Transition Town Movement, which focuses on building resilience in local communities).

The psychological defence of believing scientists and politicians will work it out is reinforced by evidence that, in some ways, thing *are* getting better. For the data scientist Hannah Richie, optimism lies in humanity's proven capacity to solve its problems, for example, in mending the ozone layer. And there seem to be grounds for hope: CO_2 emissions reached their peak in 2012. In Africa, 55 per cent of energy consumption came from green sources in 2021. Here in the UK, the use of solar and wind power has just outstripped coal (Blakeley, 2025). Communities are trying to take control of their energy supplies. An estimated 2 million people were involved in 23,000 community energy projects as of 2020–2021 (National Energy Foundation, 2024). Alongside the escalating use of fossil fuels in the Global North there is also a solar revolution around the world. There are strong signs that we could fix this if there was the coordination and political will to do so – to make real the agreements made at the UN Climate Change Conferences.

The psychological relevance of these simultaneous and opposite trajectories of action and inaction (however unequal) is that they set up a tension between despair and hope. We may feel both, producing a chaotic emotional response to the crisis that mirrors the crisis itself. Engaging with climate change is to enter a deeply troubling arena of conflicting narratives that are inflected with ideology and psychological defence. Awareness of the ecological crisis is a trauma, and all the more so without any coherent sense of its consequences.

The American child psychiatrist and attachment theorist Dan Siegel and his colleagues found an unpublished paper by Bowlby, which suggested that,

> an organism that experiences fear that disrupts the attachment system can be anticipated to suffer from "traumatic difficulty in cortical incompatibility of sense data."[3] The trauma results in the components of the attachment system – attention, expectation, affect, and behavior – coming apart from one another.
>
> (Reisz, Duschinsky, & Siegel, 2018)

For many, the core of climate-awareness trauma is a sense of confusion, powerlessness and entrapment. Maggie Turp (chapter 4) offers some sense of how profound the terror can be, not just for those who already live in disrupted ecologies but also those who anticipate global devastation. She refers to the concept of "psychic skin" (Esther Bick's term for the containment provided by the mother that gives the baby a sense of self). When this fails, the infant feels that they have "to hold themselves together, to avoid the terrible fate" – to quote Bick's words – of "liquefying and leaking away" (p. 46). To lose Earth as a living ecosystem is to lose all that contains and gives us life. It is terrifying. Many of us will look away and carry on as usual. Others will resort to defensive fantasies of self-sufficiency and survivalism (Jeremy Holmes Chapter 1, p. 4). Others will take up an aggressive, pro-fossil fuel posture, not least Donald Trump, who is calling climate change "the greatest con job ever perpetrated on the world" while the US produces 22 million barrels of oil a day.[4]

James Angel (Chapter 2) suggests that psychoanalysis can offer a more nuanced narrative for understanding abdication or denial of climate change than the idea that some people are simply callous, selfish and greedy. Freud, in particular, was fascinated by destructive impulses that are out of conscious awareness, and he considered whether his concept of the death drive might help to explain our ever-escalating momentum towards global destruction. Yet, however cleverly the mind can compartmentalise or be unconsciously manipulated into climate denial by vested interest groups, surely every adult knows that there's a problem for which we are all responsible. I recall a conversation with a friend who fully understands this but wants to spend his retirement flying to far-off places. He believes carbon emissions from his private travel are a drop in the ocean, and he shouldn't have to make a meaningless individual sacrifice. I frequently make deals with myself to have what I want, such as the idea that giving up meat entitles me to tropical fruits even if their cultivation devastates local ecosystems, or buying jeans when I'm aware of the ecological damage caused by their production. The psychoanalyst Sally Weintrobe, who has written extensively about attitudes underpinning climate change and is widely quoted in this book, says that this attitude of "exceptionalism has greatly escalated harm to people and planet during the era of neoliberalism, an ideology-driven deregulated form of capitalism that has gained ascendancy over the past forty odd years" (Weintrobe, 2014). One of capitalism's dominant narratives is that we need and deserve gadgets, foods and experiences that are better, cheaper and quicker. The more we demand environmentally expensive products and services, the better capitalism does. And, for those with money, consumerism endows a sense of security and status.

But is this an attachment problem or an ethical one? Even if politicians and business giants continue to use Earth as a limitless resource, the individual or community can choose between being a bystander and becoming an active protector of Earth. And there is a substantial demographic across the world that is ecologically engaged. What do we know about their personality types or early attachment ties that make them so?

To track down answers, it's interesting to look at some studies. A Norwegian group in 2021 at the University of Oslo concluded that attachment security leads to a greater "perceived responsibility towards anthropogenic climate change and increased empathy for others." This group of researchers propose that climate change mitigation is best served by "psychological structures related to human interconnection": secure types tend to seek proximity to others in the face of threat, while those who are insecurely attached "tend to remain focused on their own distress." They also found evidence that attachment security leads to increasing levels of empathy, trust and helping behaviour (Nisa et al., 2021).

But there are also those with insecure or traumatic childhoods who resonate deeply with the other-than-human, who express a profound sense of ethical responsibility towards nature and relate to it as their secure base. An American study in 2022 found that "trauma as a child is a positive predictor of civic engagement and green behavior" (Raja & Carrico, 2022). Researchers have also found that those identified as Highly Sensitive People have a strong sense of connectedness with "nature" and protectiveness towards animals (Cangilla, 2023). A complex picture emerges here, showing that caring for the ailing environment involves those with both secure and insecure early relationships. This is barely surprising given the magnitude of the crisis we face.

While it seems that humanity is losing the battle to mitigate climate change with sufficient speed, there are also hopeful signs of a counterculture of eco-sensibility emerging in response that is shifting Western culture towards resonance with the ecosystem. In his 1969 paper "Ecopsychology: Eight principles,"[5] the late academic and author Theodore Roszak wrote that "the core of the mind is the ecological unconscious" – that "the ecological ego matures toward a sense of ethical responsibility to the planet that is as vividly experienced as our ethical responsibility to other people" (Roszak n.d.). At an individual level this shift shows up in many ways: in choices about how to source food, travel, vote or heat our homes. At a community level, environmental campaigns and Green Party meetings both express and foster eco-sensibility. In Climate Cafes, where it's safe to share feelings about the crisis, or in Green faith groups, people are forging a synthesis of ecological responsibility and community that is based on cooperation and attachment.

Eco-sensibility is also stimulating an abundance of creativity. In her chapter (Chapter 4), Maggie Turp, offering an extensive list of titles and her critical analysis of these, discusses the emergence of *cli-fi* (climate fiction) as a recognised genre that helps us to find echoes of our troubled feelings in the imagined lives of others. These books, she says, can help us to "hold onto our capacity to feel," to make sense of things – to ease us into a new story about the world. Not only is there a flowering of ecological fiction, but there are also numerous environmental poets and songwriters. There are visual artists who use found objects, sustainable materials, or create ephemeral pieces that return to the earth. In her chapter, Catherine Jacobs reflects upon the power of art to engage with the unknown, just as we must when we unflinchingly look at the ecological crisis.

Not least, the psychotherapy field is becoming rapidly enriched with approaches, trainings and texts that are grounded in eco-sensibility, from courses in ecopsychology, to equine, canine, horticultural, wilderness and environmental arts therapies.[6] While evolving from distinct theoretical foundations, these are connected by the desire to introduce eco-sensibility into the professional practice of helping people to heal relational trauma. And, critically, these trainings do so by bringing people together into new attachment groups. I know from my own experience that the most uplifting gatherings I have ever attended are ecopsychology conferences – places where I have been momentarily able to feel the greatest affinity with others.

In the therapy consulting room, ecopsychotherapy embraces the other-than-human. It does not pathologise eco-anxiety, depression or rage, but gives space – both inner and outer – to the ecological maturity of these feelings. Its approach could not be better expressed than by Theodore Roszak: "For ecopsychology as for other therapies, the crucial stage of development is the life of the child" – just as it is in Attachment Theory. Roszak (n.d.) continues:

> Ecopsychology seeks to recover the child's innately animistic quality of experience in functionally "sane" adults. To do this, it turns to many sources, among them traditional healing techniques of primary people, nature mysticism as expressed in religion and art, the experience of wilderness, the insights of Deep Ecology.

Ecopsychotherapy and attachment-based psychotherapy are perhaps most closely aligned through the concept of the relational. Both look at where something has gone wrong between people. Ecopsychotherapy simply extends that to the space between people *and the environment*.

As the authors here have shown in so many ways, we are all searching for connections between ideas and between people that will empower us in protecting and contributing to the unique, delicate ecosystem that is Earth – in seeing it as a subject which requires respect, as a mystery, as a place of exceptional beauty and our only home. Were Bowlby still with us, as one of the great connectors, he would surely endorse this innovative book.

Notes

1 See www.rootsandshoots.org.uk/about-us
2 A note here about vocabulary: the widely used word "nature" often occurs in this book. However, it is increasingly seen as a misnomer, used to describe that which is not human or humanly made, and replaced by the term "other-than-human."
3 (PP/BOW/H.10), notes from a file tagged "Theory of Defence 1960–1963."
4 At the UN General Assembly, September 2025.
5 Also author of *The Making of a Counterculture* (1969)
6 See a list here: https://ecopsychology.org.uk/uk-training/

References

Blakeley, G. (2025). How the rich world lost the race to renewable energy. *Substack*, 13 October 2025. https://graceblakeley.substack.com/p/how-the-rich-world-lost-the-race

Cangilla, Lori L. (2023). HSPs are more connected to nature, study finds. *Highly Sensitive Refuge*, 26 July 2023. https://highlysensitiverefuge.com/highly-sensitive-people-are-more-connected-to-nature/

Miguel-Tomé, S., & Llinás, R.R. (2021). Broadening the definition of a nervous system to better understand the evolution of plants and animals. *Plant Signaling & Behavior.* 16 (10): 1927562. https://doi.org/10.1080/15592324.2021.1927562

National Energy Foundation (2024). Community energy: A sphere of innovation. *NEF Blog, November* 13, 2024. https://nef.org.uk/community-energy-a-sphere-of-innovation/

Nisa, C.F., Bélanger, J.J. Schumpe, B.M. & Sasin, E.M. (2021). Secure human attachment can promote support for climate change mitigation. *Proc. Natl. Acad. Sci. U.S.A.* 118 (37): e2101046118. https://doi.org/10.1073/pnas.2101046118

Panksepp, J., & Biven L. (2012). *The Archaeology of Mind: Neuroevolutionary Origins of Human Emotion.* W. W. Norton.

Pardikar, R. (2020). Global north is responsible for 92% of excess emissions. *EOS*, 28 October 2020. https://eos.org/articles/global-north-is-responsible-for-92-of-excess-emissions

Raja, U. S., & Carrico, A. R. (2022). Childhood trauma and other formative life experiences predict environmental engagement. *Scientific Reports*, 12(1), 20756. https://doi.org/10.1038/s41598-022-24517-7

Reisz, S., Duschinsky, R., & Siegel, D. J. (2018). Disorganized attachment and defense: Exploring John Bowlby's unpublished reflections. *Attachment & Human Development.* 20(2): 107–134. https://doi.org/10.1080/14616734.2017.1380055

Roszak, T. (1969). *The Making of a Counterculture*. New York: Doubleday & Co.

Roszak, T. (n.d.). Ecopsychology: Eight principles. *Ecotherapy Oxford.* https://ecotherapyoxford.co.uk/what-is-ecotherapy/ecopsychology-eight-principles/#:~:text=1.,and%20family%2C%20person%20and%20society

Weintrobe, S. (2014). We live in a culture of uncare. www.sallyweintrobe.com/blog/29-nov-2014-culture-of-uncare/

Weir, K. (2025). Nurtured by nature. *Jnl of the American Psychological Association.* 51(3). www.apa.org/monitor/2020/04/nurtured-nature

Preface

Linda Cundy

This book is based on presentations made to the 27th John Bowlby Memorial Conference in 2024 on the theme of Attachment, Climate Crisis and the Natural World, with supplementary chapters contributed by others who were not speakers at that event. There have been many publications and conferences about ecopsychology previously, but this is the first to look at these critical issues solely through the lens of attachment. The authors reflect on our current relationship with the natural world from the perspective of Bowlby's major themes, the nature and primacy of the attachment instinct for survival, the secure base, separation anxiety, maternal deprivation, loss and mourning. The therapeutic value of renewing our relationship with, and respect for, the natural world we are part of is explored, as is working with psychotherapy clients who are troubled by climate change, loss of biodiversity, suffering caused around the world by continued reckless plundering of natural resources, and so on.

Why Attachment and the environment?

If Bowlby were alive today, I feel sure that the ecological emergency would be a matter of deep personal and professional concern to him. His thinking was greatly influenced by Darwin – indeed, he wrote a biography of the great naturalist (Bowlby, 1990) – and Attachment Theory has incorporated findings from ethology, the study of species in their natural environments. We humans are animals living in and adapted to an environmental ecosystem that is now under serious threat. Biographer Susan van Dijken noted Bowlby's love of the natural world. At the progressive Priory Gate School where he taught for a time as a young man, "nature was viewed as an educator," one of the attractions for the young Bowlby. This rather reminds me of some of the projects Roger Duncan (Chapter 3) has been involved with at Ruskin Mill Education.

Bowlby grounded his theory not only in an evolutionary context but in historical, political, and economic realities. As such, it has much to offer when reflecting on the effect of our behaviours on the environment, the devastating impact of global warming and ecological degradation on human and other-than-human experience, and the need for effective strategies that engage and unite individuals to work together. He was a political animal with a desire to "improve the community as a whole" (Van Dijken, 1998, p. 46) through promoting secure attachment between children and their parents and also within communities. A democratic socialist with a profound commitment to social responsibility, he was a close friend of Labour Party politician and political philosopher Evan Durbin (Mayhew, 2006). Both became early members of the New Fabian Research Bureau, committed to social justice through psychologically informed policies. He was considered radical, certainly controversial in psychoanalytic circles at that time, for his insistence that we need to take a person's social, economic, environmental and relational realities into consideration to understand their inner world. His view was that we should observe infants and young children, note the kinds of experiences – such as separations from their parents and familiar environments – that lead to distress and long-term personality changes, then legislate to minimise their exposure to such "pathogens."

Bowlby would surely have been alert to the impact of increasingly frequent and severe climate events and eco-anxiety on caregivers and communities, and the repercussions for childcare and child development. I believe he would also have been keenly aware of the detrimental impact of air pollution on children's health and been actively engaged at a political level to promote policies that limit carbon emissions and restore habitats.

Addressing the ecological and climate emergency is about far more than reducing carbon emissions – we also need to ensure a more equitable world made up of secure societies with cultures of caring for, encouraging, and valuing their citizens. Colonialist greed and the consumerist demands and dominance of the Global North have been major factors in the ecological crisis we have created, and that is disproportionately impacting those who have contributed least to the greenhouse effect. Subsequent research demonstrates that secure attachment provides the best conditions for our innate altruism and co-operative nature – as a profoundly social species – to flourish and to benefit the world, while insecure attachment can potentially lead to selfishness, greed, rivalry, aggression and sometimes cruelty, all motivated by psychological defences (Music, 2014).

John Bowlby was buried on the Waternish Peninsula on the beautiful and rugged Hebridean Isle of Skye, where his family had spent their holidays surrounded by majestic scenery and wildlife – red deer, grey seals, sea otters, dolphins and eagles – and where he felt at peace.

***

The Bowlby Centre's founder John Southgate was supervised by Bowlby and was also a politically engaged radical, author of *The Barefoot Psychoanalyst*

(1989), and Consultant to the Institute for Social Inventions (ISI) and its Council for Posterity. The aim of the ISI was to stimulate and promote "new and imaginative solutions to social problems."

This Council for Posterity's draft document (below), produced with Southgate's input, is a stark reminder that concerns about the impact of human activity on the planet are far from new, and it lays out a far-reaching vision and philosophy to address climate justice and human rights.

Universal Declaration of the Rights of Posterity (draft, July 1989)

Whereas our descendants' generations, although likely to prove far more numerous than ourselves, have no voice or vote in any of our political systems, we therefore declare that all the generations yet to come, for as far into the future as human life survives on this planet, are entitled by natural justice to the following rights, which those alive today have a duty to respect and uphold.

- i the right to inherit a planet that has been treated by its human caretakers with the utmost respect, and more particularly
- ii the right to inherit a planet with an atmosphere undamaged by human activity, with safeguards for the planet's climate, such as the rainforests still intact
- iii the right to inherit a planet with an undiminished diversity of species
- iv the right to inherit adequately maintained reserves of mineral, fuel and other resources
- v the right to be born into a world of human-scale societies, unravaged by population excess
- vi the right to inherit a world unmarred by nuclear or chemical wars, incidents or wastes
- vii the right to expect that previous generations will have cleaned up their pollution, repaired their damaged environments on land, sea and in the air, including outer space and protected places of natural beauty
- viii the right to inherit political, legal, technological and social systems that respect individual human rights, such as those laid out in the UN Declaration of Human Rights
- ix the right to inherit the full uncensored cultural heritage bequeathed by former generations, as expressed, for instance, in their art, history, literature, libraries and museums
- x the right for future generations to have their interests considered and, where appropriate, represented by an advocate in any present-day tribunal or assembly that is deciding courses of action that might have harmful long-term consequences

This is our ethical inheritance; 2026, the year this book is published, marks the 40th anniversary of the Bowlby Centre. It is appropriate and timely that we are engaged in exploring our relationship with the natural world and the impact of ecological and climate crises from an attachment perspective.

***

Adverse Childhood Experiences

Bowlby was insistent that his theory be backed by sound evidence. In keeping with his motivation to promote secure attachment between children and their parents and within communities, a huge international research project has been conducted over the past 30 years or so years investigating the sequelae of Adverse Childhood Experiences. This clearly demonstrates that people who have grown up in environments with multiple stressors are substantially more likely than others to experience all physical and mental health problems, as well as social disadvantages.

The ACEs research project was initially developed by Vincent Felitti and his team. He had set up an obesity treatment programme funded by the Kaiser Permanente Health Insurance provider in the 1980s to treat morbidly obese participants. Through the often life-saving programme he devised, patients' weight loss over the course of 12 months was dramatic – but some immediately regained those kilos and more at the end of treatment. The ACEs research uncovered a connection between these people's self-destructive relationship with food and multiple harmful factors in their early lives; parents and family members who had mental health problems, criminal records or were addicts, living in an atmosphere of domestic violence, physical, emotional, or sexual abuse, bullying, neglect, and so on (Felitti, 2002). Depression, self-loathing and poverty are not the best foundations for feeding oneself thoughtfully. High-calorie, low-nutritional-value, cheap ultra-processed foods are then often favoured and can become addictive, and parents caught in this cycle of deprivation and hopelessness are unlikely to feed their children healthily or provide an atmosphere of secure attachment.

We now know much more about the impact of ultra-processed foods on bodies and minds. However, UPFs also have a serious and significant impact on the environment. They are;

> responsible for significant diet-related energy, greenhouse gas emissions, land use, water use, energy consumption, land degradation, biodiversity loss, use of fertilisers, herbicides, eutrophication, food loss and waste, and packaging waste, occurring in multiple stages of the supply chain.
>
> (Anastasiou et al., 2022)

Felitti's original research led to the development of the ACEs questionnaire, focusing on ten potential childhood (before the age of 18) stressors. This has subsequently been expanded into the Adverse Childhood Experiences International Questionnaire, the ACE-IQ – now including experiences of forced marriage, war, torture, organised violent crime, gang warfare and political or ethnic conflicts. Additional questions include "Were you forced to go and live in another place due to any of these events?" and "Did you experience the deliberate destruction of your home due to any of these events?" Doppelt (2019) and Thapa et al. (2025) make a convincing argument for the climate and eco-crisis to be added to the ACEs list of criteria [E-ACEs – Environmentally-driven Adverse Childhood Experiences]. Extreme climate events can lead to parental stress, loss of home, separation from family, loss of livelihood, food insecurity, disrupted education, deterioration in living conditions, and so on.

In the editorial of a special edition of the Bowlby Centre journal, guest editor Simon Partridge issued a rallying call: "I suggest that those … involved in succouring human well-being, whether professionals, survivors, or community enablers, now gather under the growing banner of the ACEs movement [that] reiterates the progressive intentions of Bowlby and his colleagues" (Partridge, 2021, xi). The social and political roots of ACES are inextricably linked with the environment. If parents, families, neighbourhoods and communities are stressed, they are likely to be more reactive. Rising temperatures may well lead to rising tempers.

According to the Climate Trauma Project in Canada, recent heat waves and droughts there have led to an increase in aggression and violence, as well as depression and possibly addictions. "Slashing greenhouse gas emissions, and conserving forests and other ecological systems that sequester carbon, are vital to reducing ACEs and the violence that can result" (Doppelt, 2019).

***

Mental health

There is clear evidence that mental health is suffering globally due to the climate crisis. A recent report from the Mental Health Foundation UK notes both direct and indirect effects. The first refers to those people around the globe who have experienced extreme weather conditions and natural disasters – drought, floods, heatwaves, wildfires, hurricanes – even volcanic activity may be influenced by global warming. As well as PTSD, people caught up in these kinds of events may have lost loved ones, homes, livelihoods, communities, precious possessions, pets and livestock. Many are permanently displaced – effectively homeless – while others may eventually return but never feel secure again in their homes. Depression, chronic grief, addictions and suicidal ideation are long-term potential sequelae. The report notes that those living in poverty, especially women and children, are most vulnerable. In recent news here in the UK, heavy rains causing flooding have led to loss of crops and livestock – and, thus, income – for farmers, having a significant impact on the mental health of individuals who already suffer more depression, anxiety and suicidality than the general population. These events are likely to become more frequent and more severe in the coming years.

Indirect effects refer to the sense of anxiety, fear, rage, guilt, helplessness and hopelessness that may be felt by those who feel overwhelmed by thoughts of the climate crisis. This is often referred to as eco-anxiety or solastalgia. It seems that children and young people are particularly vulnerable. A survey conducted by the University of Bath suggests that around four in ten young people fear having children as a result of climate change and ecological emergency.

Eco-anxiety

I want to tell you just a few recent vignettes from my therapy practice of the indirect impact of the environmental emergency on mental health.

> Since the death of his partner, one man has become a hoarder. We have explored many of the meanings and functions of this symptom – objects as companions, identification with "unwanted" items, to keep others away, and so on (Cundy, 2020). But a significant factor is his concern over recycling. How can he avoid adding to landfill? He has a strong and pervasive sense of guilt about this. Can objects be repurposed? Can he find a new home for various bits of equipment he no longer needs? What can go to charity shops? What can be recycled, and where? These decisions paralyse him.
>
> The mother of a child on the autism spectrum is another patient. Her daughter became very anxious and dysregulated after a school lesson about climate change. A remarkable mother, she was able to soothe her daughter by telling her about Greta Thunberg and her autism spectrum "disorder." She helped the girl reflect that neurodivergent individuals may have gifts that contribute to finding solutions.
>
> Another client is a young man working in the public sector. He is deeply engaged politically but is overwhelmed and depressed by the daily news coverage, drowning under a tidal wave of horror stories, petitions, calls for donations that fill his inbox every day. He is struggling to focus on work and feels despair, asking, "What's the point of bringing children into this world?" This is affecting his relationship with his partner, with frequent conflict over day-to-day household issues – use of lighting, water, cleaning products, recycling, heating. His relationship is at risk.
>
> An older woman client has reported having panic attacks and nightmares. She was active in the Campaign for Nuclear Disarmament (CND) movement and involved with the Greenham Common peace camp in the

1980s. This was an era of impending nuclear threat – MAD (mutually assured destruction). Her recent symptoms conflate nuclear devastation with ecological collapse. She fears for her grandchildren.

***

Jeremy Holmes (Chapter 1), our conference Key Note speaker, argued back in 1996 for Attachment Theory as "a secure base for policy"; that "connection, watchfulness and ecological awareness … could be the starting points for a psychologically informed twenty-first-century politics. Security rests on recognising and valuing our connections, locally, nationally, internationally and interspecifically" (Holmes, 1996, p. 27).

Authors here consider the significance of Attachment Theory now, 30 years on, for understanding the climate crisis, but also propose what we can do, individually and in the various communities we belong to, to keep up the pressure, to make changes, to have our voices heard – and to hold onto our humanity in the face of the catastrophic impact of global warming. Heglar recommends, "do what you're good at. And do your best" (2023, p. 25). Scientists, activists and politicians can all use their skills to be part of making a difference. As Maggie Turp (Chapter 4) and Catherine Jacobs (Chapter 5) highlight, artists and writers can use their creativity as a platform to unite people with the common aim of environmental activism. Psychotherapists can play their part too. In the chapters that follow, authors reflect on how to address climate anxiety in the consulting room, with James Angel making the case for a "politically engaged psychotherapy" (Chapter 2), or outside it with nature-based therapy practices (Roger Duncan, Chapter 3). The final chapter (Linda Cundy et al., Chapter 6) explores the importance of landscape in one's identity and in healing from trauma, while Jeremy Holmes (Chapter 1) provides a philosophical and theoretical foundation to help us mentalise in the face of growing stress and trauma.

The function of attachment is to ensure survival; that encompasses not only the survival of individuals and communities but of entire ecosystems.

***

References

Anastasiou, K., Baker, P., Hadjikakou, M., & Hendry, G. (2022). A conceptual framework for understanding the environmental impacts of ultra-processed foods and implications for sustainable food systems. *Journal of Cleaner Production*. 368(4). https://doi.org/10.1016/j.jclepro.2022.133155

Bowlby, J. (1990). *Charles Darwin: A New Life*. London: W.W. Norton.

Cundy, L. (2020). A brief attachment perspective on hoarding. *Attachment: New Directions in Psychotherapy and Relational Psychoanalysis.* 14(2): 168–170.

Doppelt, B. (2019). Climate Change as ACE. The Climate Trauma Project. www.pacesconnection.com/g/the-climate-trauma-project/blog/climate-change-as-ace (accessed 13 August 2025).

Felitti, V.J. (2002). The relation between adverse childhood experiences and adult health: Turning gold into lead. *Permanente Journal.* 6(1): 44–47.

Heglar, M.A. (2023). Here's where you come in. In Solnit, R. & Lutunatabua, T.Y. (Eds.) *Not Too Late: Changing the Climate Story from Despair to Possibility* (pp. 19–27). Chicago, IL: Haymarket.

Holmes, J. (1996). Attachment Theory as a secure base for policy? In Kraemer, S. & Roberts, J. (Eds.), *The Politics of Attachment: Towards a Secure Society* (pp. 27–42). London: Free Association.

Mayhew, B. (2006). Between love and aggression: The politics of John Bowlby. *History of the Human Sciences.* 19(4): 19–35.

Music, G. (2014). *The Good Life: Wellbeing and the New Science of Altruism, Selfishness and Immorality*. Hove, East Sussex: Routledge.

Partridge, S. (2021). Editorial. *Attachment: New Directions in Psychotherapy and Relational Psychoanalysis.* 15: xi.

Southgate, J. (1989). *The Barefoot Psychoanalyst: Illustrated Manual of Self-Help Therapy*. London: Gale Centre.

Thapa, S., Giri, S., & Ross, A.G. (2025). Extreme weather and climate-related adverse childhood experiences: A humanitarian crisis of the 21st century. https://doi.org/10.20944/preprints202505.2296.v1.

Van Dijken, S. (1998). *John Bowlby: His Early Life. A Biographical Journey into the Roots of Attachment Theory*. London: Free Association Books.

Chapter 1

Towards an ecology of attachment

Jeremy Holmes

Albert Einstein and Sigmund Freud were both ardent pacifists, deeply opposed to war as a means of resolving disputes. But they were also both realists and acknowledged that human history can, without much distortion, be charted as one long history of warfare. With Fascism looming, in 1932 Einstein wrote to Freud asking him for his views on human aggression and violence, and how future wars might be avoided (Leith, 2000).

Freud's response was not entirely pessimistic. While he saw aggression as a necessary part of the human psyche, this was counterbalanced by his celebration of Eros, which, alongside sexuality, fosters collaboration, community and mutual identification. In addition, through his theory of the superego he saw conscience as an internalisation and therefore mitigator of aggression.

Attachment Theory has been criticised for its downplaying of the role of sexuality, but if we take Eros as encompassing emotional, physical and social connectedness, then we can think of Attachment Theory as extending and developing Freud's concept of Eros. As Bowlby (1971) put it, the everyday word for attachment is love.

My starting point therefore is love: love of nature, love of the world as our ultimate secure base, and love of each other as a species.

In this broad-brush chapter, based on my conference presentation, I shall try to show how attachment ideas can help us understand our varying psychological responses to climate change and point to ways in which, through concerted action, it can be mitigated.

***

Disavowal

A key concept developed in the work of Sally Weintrobe (2021) is *disavowal*. Freud's term was *Verleugnung*, which commonly translates as denial or "disownment" but was called by Freud's translator Strachey – in his attempt to endow psychoanalysis with a quasi-scientific status and terminology – "disavowal." Freud wanted to capture the defence mechanism by which, when

DOI: 10.4324/9781003733942-1

faced with a traumatic perception or memory, one simultaneously acknowledges and rejects a fact or facts. To give a painful personal example, when working as a psychiatrist, a GP rang me with the news that one of my patients had, very unexpectedly, ended her life. My initial response was to say to myself, "Oh no, this *can't* be true. The GP (who was something of a joker) must be trying to fool me." But of course, I knew full well the horrible reality, and my disavowal was no more than a feeble attempt to keep my feelings of sadness and guilt and failure at bay.

Weintrobe suggests that denial and disavowal encapsulate the ambivalent attitude that most of us have in relation to climate change. We continue with life as normal – getting on planes, turning up the heating or air conditioning, guzzling meat, driving our SUVs. Yet at the same time, we know that the planet is heating, that people (although not "us") are dying as a result today and in the future, and that our children and grandchildren will inevitably be adversely affected.

Thus does disavowal capture prevalent attitudes to climate change. But in itself it explains nothing. Describing is only a first step towards understanding. We need to ask, what is the nature of this disavowal, where does it come from, how can it be overcome, and what might constitute effective action to mitigate or reverse its effects?

Attachment Theory

For that we need to return to the trauma which triggers the disavowal. Trauma literally means a piercing. What's traumatic about trauma is that it cuts through the models of ourselves and our lives with which we shield ourselves from uncertainty and danger. Seeing and thinking about what is happening to the natural environment – either through direct experience, or through media reports of weather extremes, flooding, rising temperatures, melting icebergs and a brought-to-a-halt North Atlantic Current – constitutes the climatic traumata with which we are daily confronted.

What can the science of attachment bring to confronting this trauma and add to the urgently needed conversation about climate change and the natural world? Attachment is both a developmental theory and a therapeutic modality (Holmes & Slade, 2017). It also provides an implicit set of principles for living, thinking and acting (Mikulincer & Shaver, 2007; Holmes, 2024a). I've come rather late to this idea, but I've found that thinking in attachment terms provides a framework with which to lead my everyday life as well as my therapeutic work. What are these principles, and how can we apply them to the climate crisis?

The first point is the quasi-tautological realisation that security is fundamental to survival. Without our immune system, we would not survive as living creatures. Attachment Theory tells us that evolution has also endowed us with an analogous set of behavioural and psychological structures which enable us to survive psychologically, the basis of which is secure attachment.

Second, this security is relational. Our physical and psychological safety, from cradle to grave, depends on our relationship with others, extending from our primary attachment figures, through Dunbar's (1993) "social brain" number of 150 contacts, to the wider community and context of which we are a part. We have to think about the systemic field *between* individuals rather than, as in the psychoanalytic world, to focus exclusively on the individual and their inner world, important though that project still is. Seen this way, the attachment model conceives us as *part of, and inseparable from nature, rather than distinct from it*. Then perspective invites us to confront what we are doing to the planet, and what it is doing, as it were, back to us, and begin to "own" climate change and the reversals and mitigations needed to contain it.

Third, loss is the inevitable reverse side of attachment. We are familiar with the ways in which our loss-traumatised patients avoid close relationships so as to avoid repeating the unbearable pain and disruption of their previous losses. Comparably, if we are not attached in an intimate way to the planet, through love of the natural world, awe at its immensity and incomprehensibility, and lack cognitive understanding of the networks of interconnection that sustain us, we are likely to turn a blind eye to the catastrophic losses implicit in unmitigated climate change.

A fourth idea integral to the attachment worldview is the importance of self-reflection, or mentalising. This is the recursive capacity to look at oneself looking at the world, a capacity that's associated with mental health, and conversely, whose lack is a feature of psychological illness, especially in borderline/emotionally unstable personality disorder, is a possible developmental outcome of disorganised attachment in infancy (Mikulincer & Shaver, 2007).

Disorganised attachment

The attachment dynamic ensures that, when threatened, we turn to our secure base for comfort and safety. In disorganised attachment, if we accept Mary Main's conceptualisation (Duschinsky, 2015), the child is therefore faced with an approach-avoidance dilemma. The caregiver is in some way threatening – through neglect, or sexual and physical or psychological violence; according to the attachment dynamic, this triggers the child to seek out their secure base for safety. But if the secure base is the very source of threat, what then ensues? The disorganised child engages in various attempts at self-soothing – rocking backwards and forwards, head banging, curling up in a ball, etc. Developmental continuity means that in early adolescence and adulthood, these may transmute into "pathological" self-soothing strategies such as deliberate self-harm, substance abuse, risky sex, etc.

I want to argue that this dynamic encapsulates the ways in which, through disavowal, we're in a state of disorganised attachment *vis-à-vis* the Earth. Mother Earth is our secure base. But climate change means that our hitherto secure base is now the very source of threat. If that secure base is too hot, if there are floods, droughts, wildfires, rising sea levels, then the home to which we would turn becomes the very source of the threat that's causing us to seek it out. We no longer trust the world which supports us – the non-human life, animate and inanimate, that surrounds and sustains us. And just as the disorganised child turns to self-defeating ways of evading this insoluble dilemma, we see comparable phenomena as we face the inexorable rise in CO_2 emissions, unabated after 27 years of COP meetings (Helm, 2020).

At an individual level, attempts to solve the disorganised attachment dilemma include fantasised self-sufficiency, various attempts at comfort through substance abuse or deliberate self-harm, which often mobilise a modicum of security, real or imagined.

The impact of climate change evokes comparable defensive strategies. Survivalism is one – retreating into a pseudo-self-sufficient enclave while the world burns. Relocating to Mars, no less. Retreating into a virtual world where danger and chaos can be gamed rather than lived is another. The manifestations of climate change disavowal are multifarious; all can be seen as vain attempts to square the disorganised attachment circle.

Flux is the nature of our existence. Everything changes; nothing stays the same. Securely attached individuals move easily and freely from one state of mind and activity to another. The insecurely attached find such transitions difficult and threatening. They cling to established top-down cognitive and behavioural pathways, however maladaptive (Holmes, 2020). The Italian Marxist Antonio Gramsci famously conjured up the image of the "pregnant widow" (Gramsci, 1971), in which a crisis arises when the old is dying, and the new cannot be born. In this interregnum, he says, "a great variety of morbid symptoms appear."

We are in a stage of great transition from one energy and power source to another. But we cling insecurely to the old ways, fearful of what an oil and carbon-free world would look like. This would be one in which we would drive less and walk or cycle more, eat less meat, value nature, equalise gross income disparities both within nations and between them, and find ways to reconcile our differences without energy-catastrophic warfare. At present, we appear further and further away from giving birth, however pregnant with good ideas – geoengineering, carbon capture, microbial meat production – to a new world order. The philosopher Peter Gray (2024) is highly critical, contemptuous even, of this liberal wish-list. But the Musk Mars-phantasy is even more unrealistic. Both, perhaps, represent Gramscian "morbid symptoms" awaiting alleviation.

Peter Marris

One of the relatively few sociologists who have incorporated Attachment Theory into their thinking was Peter Marris (1996). His focus was on how, as societies rather than individuals, we manage uncertainty. Given that we cannot know what the future holds and, faced with an ever-changing and unpredictable world, societies need a security system if they are to survive (which, in the long run, of course, none do). Marris' insight was that the rich and powerful maintain their own security by *exporting insecurity* to the poor and less powerful. Thus, in our contemporary world, in the so-called gig economy large corporations transfer the insecurities of the market onto individuals who lack employment rights, pension provision, healthcare benefits, holiday pay, etc. – everything that corporations and their comfortable employees take for granted. Insecurity is exported into an underclass, thereby amplifying their insecurity.

Marris linked the psychology of attachment with the sociology of inequality. Wilkinson and Pickett (2009) later charted the perversity of this process, showing how Marrisian projection of insecurity boomerangs back on the rich. Looking at indices of social disorder – rates of suicide, substance abuse, crime statistics – they found that these strongly correlate with the extent of the gap between rich and poor. Their findings applied both between the 21 richest countries and also within the US between States. While the impact of this structural inequality falls differentially on the poor, it also affects the rich. In a vicious cycle, the more insecure the rich feel, the more they cling to their flimsy privileges, and the greater their levels of anxiety and depression, and the more they try to evade insecurity by getting others to suffer it in their stead.

This analysis can be applied to climate change. From a global perspective, the North is responsible for about 90 per cent of CO_2 emissions. Meanwhile, the impact of the greenhouse effect is most powerfully felt in the South. Countries which are most impacted by, and least well equipped to deal with the consequences of climate change – extreme weather events, rising sea levels, etc. – are those that have contributed least to its rise. Hickel (2021) calls this *atmospheric colonisation*.

The tragedy of the commons

While it is hard to deny that the Global North is exploiting and impoverishing the Global South, it is important to take a systemic rather than simplistic and paranoid view of these processes. Consider the problem posed by the so-called "tragedy of the commons" (Hardin, 1968). If you have common land (or shared atmosphere), anyone can graze their cattle – or breathe – there. That's fine, until a few people want to put just one more sheep or cow onto that land. What was formally held in common now fosters competition for a scarce resource. The balance between individual need and common good is disrupted. And for the more powerful, need soon turns into greed, until, before you know it, we have enclosures, displacement of peasants, starvation, homelessness, migration and urban degradation.

Let's apply the tragedy of the commons principle to climate change. I – a denizen of the Global North – might think I'm free to catch a long-haul flight, cook that tasty steak, turn up my central heating or air conditioning, run my SUV round the corner to pick up the kids from school. Yes, it will pump a bit of CO_2 into the atmosphere, but it's hardly going to harm me, so let the common good go hang. And even if I did act more responsibly, individual actions – often fostered by the government lobbying the oil and gas industry as "solutions" – are only going to reduce global CO_2 levels by 10 per cent – so let "the government" sort it out.

But – and here's the tragedy – if *everyone* is saying that, then CO_2 levels will continue inexorably to rise. Our atmospheric commons is increasingly degraded. Soon, none of us will be able to breathe freely or graze our respiratory cattle.

And this takes us to another iconic problem, the prisoner's dilemma (PD; Axelrod & Hamilton, 1981). Two prisoners are offered a choice by their captors. Either you betray your fellow inmate, in which case he gets a long sentence, and you go free, or you admit joint responsibility and you both get reduced sentences. The dilemma is that your fellow sufferer is offered exactly the same bargain, and you have to second-guess how he will opt. You're faced with a choice between being very selfish, which means you may get a lot of benefit, but if your mate is equally selfish a lot of disbenefit – or being altruistic, in which you get moderate benefit, although in some versions of PD, the sum of your joint benefits exceeds that which the individual achieves.

If you take the high-risk path and you're selfish, it's very possible that your fellow prisoner may be equally selfish, in which case you *both* end up with long sentences. To mitigate climate change, we need to persuade people to work towards a common good in which all benefit, and to argue that by following individual gain, we run the risk of ending up with nothing, or worse. Ignoring climate change is a zero-sum game in the sense that the Global North's gain is the South's equal and opposite loss. On the side of cooperation is the fact that the pain of loss – eating less meat, flying less, turning down the heating – is typically greater than the pleasure of gain. In the short-term, our health benefits from more exercise and eating better; in the long term, we will be securing a future for our grandchildren.

There is a neoliberal argument which says we don't need to worry about climate change because renewable resources are much cheaper than coal, thus the market will ensure that the move from fossil fuels to renewables will happen anyway without government intervention. However, in Marris' formulation, the precondition for collaborative action is mutual trust. Concerningly, what we're seeing at the moment, particularly in North America, is a breakdown of trust – between citizens and one another; between citizens and their government; and thus between political parties, who resist the democratic bedrocks of the rule of law, the role of scientific evidence and truth, and acceptance of constitutional principles.

Elinor Ostrom

There is an answer to the tragedy of the commons, developed by Nobel Prize-winning economist Elinor Ostrom (1990). She studied collaborative situations in which resources are scarce and need to be shared. One example centres on water supply in a mountainous area of Bali, and the ways in which groups who live in different parts of the region have developed structures for preventing any one village stealing more than their share of the precious water resource. Drawing on this and other examples, she lists a number of principles by which groups can successfully collaborate and avoid the commons' tragedy. These include clear and mutually agreed boundaries, graded sanctions for misdeeds, open discussion of differences, and generally accepted methods for resolving disputes.

These principles mean that there is a "low cost," war-averse means of dispute resolution. These will be needed when planning for climate change, because when it comes to renewable energy sources such as wind turbines and solar panels there are bound to be disputes, NIMBYism and more. Applying Olstrom's principles means that competing groups are less likely to get caught up in the tragedy of the commons. Collaborative working will be built into their cultural norms. In addition, bringing Olstrom's solutions to climate change validates action at a local level and thus embodies the "think global, act local" principle, which informs ecological activism. That said, the continuing failure of COP meetings to bring down global CO_2 levels, despite lip service being paid to Olstromic principles in the Global North-South divide over climate change, is a necessary antidote to thinking that solutions are easy to come by or enduring.

A neuroscience interlude

Recently, I have tried to bring the ideas of neuroscientist Karl Friston (2018) to bear on our practice as psychotherapists (Holmes, 2020, 2024a, 2024b). I also think his model can help us think more clearly about the psychology of climate change.

Friston's idea is that the brain's continuous aim is to minimise free energy and to avoid surprise. It does this by generating "top-down" models – stories – about the world, based on the incoming "bottom-up" information provided by our senses, from within and without, enteroceptors and exteroceptors. The brain – and living creatures generally – works against the second law of thermodynamics, and so abhors chaos and disruption. In order to maintain informational homeostasis, it minimises discrepancies between input and story.

In secure attachment, caregiver and care-seeker generate a combined story about their shared affective world. The mother who comes to her crying child in the middle of the night co-creates the story of that child's pain, as she enacts her "questions": "Too hot? Too cold? Does your nappy need changing? Are you hungry and need a feed? Have you had a bad dream?" With her gestures and actions she and the child are engaging in a two-way "conversation," as she listens to and watches the child's responses: "No, it wasn't my nappy, that's not going to stop me crying, I need a feed," etc.

In insecure attachments, and especially disorganised attachment, these "conversations" either don't happen at all or fail to run smoothly. There's no interactive relational conversation, and so no shared story between caregiver and care-seeker. Neither party is in the state of "active inference" (i.e. exploring the world in order to generate more accurate models of it), which in Friston's model is the main means by which energy is "bound" in the brain.

In relation to climate change, what's needed are stories – top-down models – which can be generally agreed on and shared. Our planet is in a state of unbound energy, literally and informationally. The energy associated with climatic instability needs to be bound by an informationally aligned story – let's call it "the truth" – without which no effective action will have purchase on the problem of rising CO_2 levels and its consequences.

How does the brain typically bind bottom-up informational energy? The first step is to increase sensory precision. You hear a vaguely untoward noise in the night; the first thing to do is to sit up and listen again. Was that a picture that had fallen off the wall? A fox scrabbling after the chickens? This then leads to action – go to check on the pictures, or wake your companion and ask if s/he heard that noise?

At this point, you are working together as a team. You've recruited another to create what Friston and Frith (2015) call a "duet for one," collective active inference. Thus do we bind energy in our everyday relationships, and these are also the ways, I believe, that we have to do so in relation to climate change. We're living in an unstable world that's changing all the time. Duets for one means that we can bind free energy more effectively if there's more than one of us, in this case, a duet. But it could be a quartet or a whole orchestra or a nation or even a world. Collective free energy binding is the way forward.

Caregivers and care-seekers bind energy together. Mother and the baby bind the energy of that child's cry together into a story. "Ah, you needed a feed. Now you're fine. Now you can go back to sleep." In psychotherapy we create duets for one so that we can more effectively bind the free energy of the patient's mental pain, with which therapists vicariously and empathically resonate.

Were the "conversations" between government and the "people" secure, they would each take responsibility for listening to one another, minimising bottom-up anxiety and top-down ideology, and arriving at a consensual model sufficient to trigger effective action. In short, democracy. In relation to climate change, this might include stopping the traffic on the M25 Motorway in an effort to initiate energy minimisation rather than disavowal. In the insecure attachments with which the climate change debate is currently constrained, conversations tend to be suppressed/repressed, pass by one another, or be vacuous.

A cartoon joke that might help us here. Two young fish, A and B, pass fish C one morning. "How's the water?" C asks them and swims off. Later, A turns to B and asks, "What the hell is water?" If we're living in a fairly stable environment, then we actually mirror that environment, and no free energy is released. We meld into it. But take those fish out of water, and in an instant they will know what water is.

Planet Earth is constantly in a state of change. But in the Anthropocene the rate of change is hugely accelerated. Our mirroring of the environment is increasingly out of step with reality. We cannot rely on our internal homeostasis to keep ourselves safe. Like fish A, what the hell is CO_2, we might ask ourselves? In the Global South we are becoming fish out of water. When the external temperature reaches a humidity of 100 per cent and 35 degrees, physiological homeostasis breaks down – we must move, change or die.

Change entails, as a first step, understanding the environment we find ourselves in. When therapists work with disorganised attachment, or borderline personality disorder, a prime objective is to instil the capacity to mentalise. Mentalising, or self and other awareness, entails understanding oneself in relation to one's environment (primarily other people), including beginning to take account of the ways in which existing strategies can be maladaptive. Good therapy always entails specificity – what specific feeling did this specific instance induce in you? We take the devil-in-the-detail situations in which our patients find themselves, especially and transferentially in relation to us as their therapists, and try to work out together "what is going on." The climate change analogue here is "think global, act local": do I really need to book that flight, buy that steak, "upgrade" to that environment-degrading SUV?

Mother Nature

Climate change demands that, drawing on our attachment experience, we nurture the Earth; in return, the Earth will nurture us. Based on her anthropological and primatological expertise, Sarah Blaffer Hrdy (2009) argues that our success as a fertile species comes from the fact that we raise children collaboratively – with "alloparents" such as grandparents, siblings, cousins, neighbours, childminders, etc. This extends and, to some extent, contradicts Bowlby's and Ainsworth's depiction of mother–child monotropism. It really does take a village to raise a human child. We can extend this further by seeing Hrdy's alloparents not just as human relations, but animal relations, and non-living, locality relations. As Rachel Carson (1962/2000) put it, man is part of nature, and therefore a war against nature is inevitably a war against himself. Or in Marx's (2007) words: "to say that man's physical and mental life is linked to nature simply means that nature is linked to itself, that man is part of nature."

Insel (1922) delineates the three principal components of mental health as Person, Place and Purpose. In attachment terms, these translate as attachment to one's secure base, one's environment, and to one's ideals. I'll end by illustrating this with a famous passage from a famous novel. In Tolstoy's *War and Peace* (1873 [2010]), Prince Andrei, a cynical version of Tolstoy himself, perhaps, is in a state of terrible depression – a sad character definitely in search of a therapist. His wife, whom he didn't love (an emotion of which he considers himself incapable), has died, leaving him with a young son.

He doesn't know what to do. He has a horrible, bullying, controlling father; he's in his early 30s, but he feels that his life is over. On a trip around his estates, he sees a huge oak tree. He immediately identifies with it. Gnarled, withered, living unsymmetrically, twisted.

Spring, love, happiness – for him, these are just meaningless words. He has no faith in hopes and illusions. "The oak is right, one thousand times right. My life is finished," he says to himself. Here is this depressed man seeing his own misery reflected in the oak.

Andrei is too far from home to get back for the night, so he goes to stay with the Rostovs, his neighbouring landowners. Approaching their house, he sees a beautiful young girl, later revealed as Natasha, who takes absolutely no notice of him. After dinner, he's given a room just above Natasha, who has a girlfriend, Sonia, staying with her. It's nearly midnight, and the girls are looking out of the window, extolling the beauty of the night. Andrei hears Natasha carolling:

> It's exquisite. Sonia, Sonia, how can you sleep? Look how lovely it is. How glorious. I feel like squatting down on my heels, putting my arms around my knees this tight, flying away.

Of course, Andrei is already in love. But he has to go home the next morning. As he returns through the oak wood, all has changed. Everything is in blossom. The nightingales trill, now near, now far away. The gnarled oak tree, quite transfigured, has spread out a canopy of dark, sappy green which swoons and sways in the breeze.

> "It's the same oak," thought prince Andrei. "No. Life isn't over at thirty-one. It's not enough to know what I had. Everyone must know it too. That young girl wanted to fly away into the sky. All of them must learn to know me, and all that my life may not be to live for myself alone."

Thanks to the oak tree, Andrei has begun to mentalise his narcissism, his borderline state, his depression. With the help of nature, he begins to move to a more relational state of mind. Ruskin called the depiction of nature in art – thunderstorms, "halcyon days," in this case the advent of spring – the "pathetic fallacy." By this, he meant that nature is in itself neutral, with no interest or necessary connection to our moods. We fallaciously project our feelings into and onto nature. Ruskin's phrase implies that the link between ourselves and the weather or nature is purely metaphorical. But viewing the man–nature relationship in a systemic way reveals how states of nature can evoke certain moods in ourselves, as well as our moods determining how we view nature. The oak tree's bloom entered a positive feedback loop with Andrei's nascent love for Natasha. This applies even at a biochemical level, where ions released in woods after a rainstorm have a depression-alleviating effect via reduced cortisol levels and endorphin boost. Man sees himself in nature, but equally, nature inserts itself into the mind of man. These nature–wo/man feedback loops are cut off in depression, open in health (Stuart-Smith, 2021). Arguably, urbanism in its neoliberal form, in which nature is viewed in a consumerist quasi-colonial way, exacerbates this and contributes directly and indirectly to climate change.

My friend and colleague, the late Robert Tollemache, who, like Weintrobe, combined a passion for psychoanalysis and climate change research and activism, found, unsurprisingly perhaps, that childhood exposure to nature is associated with greater climate change awareness in adulthood, whereas lack of such exposure correlates with denial (Tollemache, 2019).

***

Getting back to nature

Freud (1916) famously claimed that science had provided three blows to our narcissism. The Copernican revolution, Darwinism and his own discovery that the unconscious, not the ego, rules our lives. A century on, these can be construed in a systemic and relational way. Our planetary system is almost certainly not unique in the universe; evolution sees species as systems rather than discrete entities; the ego is not primary but gradually emerges out of relational attachment-structured developmental processes.

As an intrinsically relational and systemic theory, attachment ideas can contribute to the climate change debate by formulating the man–nature relationship as a secure, or not so secure base and by identifying insecure states of mind such as denial, overwhelming anxiety or disavowal. Here is an extract from Kingsolver's remake of David Copperfield (2023, p. 237), which she calls Demon Copperhead, and she describes a really low-level existence.

> Impoverished people live in horrible environments, and the more impoverished, the more horrible the environment, the less exposed to nature they are. Hundreds of people pass by outside hugging their coats around them, looking at their feet, walking fast … not one thing you see is alive: engine driven steel, no morning or evening birdsong, just car horns, jackhands, mountains of steel beam constructions.

In our highly unequal Western (Global North) societies, it's the poor and the deprived and the oppressed and the colonised who are most deprived of man–nature reciprocity on which the health of both the planet and of the individuals who inhabit it depends.

My formulation of mentalising is the capacity to see oneself from the outside and others from the inside. The moon landing was seen primarily as a technological triumph (and one that entailed releasing inordinate amounts of CO_2), but it was also a mentalising triumph because, for the very first time, we found a perspective from which to see Mother Earth from the outside. The awe-struck astronauts exclaimed, "I had a feeling it's tiny, it's shiny, it's beautiful, it's home, and it's fragile." They expressed an overwhelming sense of oneness and connection. For these early astronauts, narcissistic ME was not themselves, but Mother Earth, seen from the outside.

Intrinsic to attachment is the inevitability of loss, grief, and mourning. Facing up to climate change means accepting how much has been lost through our own actions, unconscious and conscious – species depletion, climate instability and low-lying terrain flooding. Trying to escape from the pain of loss leads to tragic hatred – denial, eco anxiety, and despair. But the rhythms of loss also tell us that recovery is possible: moving towards agency, recovery of hope, and reinstatement of the lost object. I'll end therefore with T.S. Eliot from *Little Gidding*:

> There are three conditions which often look alike
> Yet differ completely, flourish in the same hedgerow:Attachment to self and to things and to persons, detachment From self and from things and from persons; and, growing between them, indifference.

Attachment to self and to things and to persons is the very essence of secure attachment. Indifference surely reflects the withdrawal from human relationships characteristic of insecure-avoidance. Which leaves us with detachment. Here, I think Eliot is grasping after the Buddhist notion of non-attachment. Non-attachment is very different from dismissing or avoidant varieties of insecure attachment. It attempts to capture a stance of attention and respect for things – people and the natural world – while at the same time accepting that nothing is fixed or permanent and all is subject to change. This stance is the very opposite of clinging, which we see in anxious patterns of attachment.

In the light of climate change, we need to find a non-attached eco-psycho-spiritual perspective (Holmes, 2024a). Our ethical imperative is to move, as with our more disturbed patients, from ego to we-go. We can learn from our work with such people: to expect turbulence, suicidality and other self-injurious states; to build epistemic trust; to work collaboratively, always "us" not "you"; to hold on to hope; to move towards mentalising.

Faced with ever-accelerating climate change, we acknowledge disempowerment. We mourn our losses. We act locally on transport, energy, food consumption, energy-conserving housing, and the environment. We make nature and green groups our secure base. And we think globally, and so try to influence government as best we can, including our democratic right to vote. I would say vote Green, but that's outside my remit, perhaps.

This takes us back to Italy, this time not to Gramsci but Dante. All the questions he asked in the 13th century – not long ago from a climate point of view – about the soul and immortality, paradise and hell flow from the unavoidable fact that every action of ours is passed on to others according to its value, good or evil, from father to son, from mother to child, from one generation to the next, and the next and the next. Dante's medium of mitigation was poetry, to tell the stories of cruelty and selfishness, of remorse, love and redemption. It's the job of therapy to face up to and help reverse the intergenerational transmission of trauma and suffering, to help people learn from their mistakes, to do better next time. Half a millennium after Dante, let's hope the same can be said about our efforts to acknowledge, own and attempt to slow down climate change. Future generations will take note.

***

References

Axelrod, Robert, & Hamilton, William D. (1981). The evolution of cooperation. *Science.* 211(4489): 1390–1396.
Bowlby, J. (1971). *Attachment*. London: Penguin
Carson, R. (1962/2000). *Silent Spring. Penguin Modern Classics*. London: Penguin.
Dunbar, R.I.M. (1993). Coevolution of neocortical size, group size and language in humans. *Behavioral and Brain Sciences.* 16(4): 681–735.
Duschinsky, R. (2015). The emergence of the Disorganised/disoriented (D) classification 1979–1982. *Hist Psychol.* 18(1): 32–46.
Freud, S. (1916). Introductory Lectures on Psycho-Analysis. In *The Standard Edition of the Complete Psychological Works of Sigmund Freud*, edited by James Strachey, vol. 15, pp. 1–240. London: Hogarth Press.
Friston, K., & Frith, C. (2015). A duet for one. *Consciousness and Cognition. 36*: 390–405.
Friston, Karl (2018). Of woodlice and men: A Bayesian account of cognition, life and consciousness. An interview with Karl Friston (by Martin Fortier & Daniel Friedman). *ALIUS Bulletin.* 2: 17–43.
Gramsci, A. (1971). *The Prison Notebooks*. Q. Hoare & G. Nowell Smith (Eds.). New York: International.
Gray. P. (2024). *The New Leviathans*. London: Penguin.
Hardin, G (1968). The tragedy of the commons. *Science.* 162(3859): 1243–1248.
Helm, D. (2020). *Net Zero: How We Stop Causing Climate Change*. London: Collins.
Hickel, J. (2021). *Less is More*. London: Penguin.
Holmes, J. (2020). *The Brain has a Mind of its Own*. London: Confer.
Holmes, J (2024a). *The Spirit of Psychotherapy*. London: Karnac
Holmes, J. (2024b). Friston, free energy, and psychoanalytic psychotherapy. *Entropy.* 26: 343. https://doi.org/10.3390/e26040343
Holmes, J. & Slade, A. (2017). *Attachment in Therapeutic Practice*. London: Sage.
Hrdy, S. (2009). *Mothers and Others: The Evolutionary Origins of Mutual Understanding*. Cambridge: Harvard University Press.
Insel, T. (1922). *Healing: Our Path from Mental Illness to Mental Health*. London: Penguin.
Kingsolver, B. (2023). *Demon Copperhead*. London: Faber.
Leith, M. (2000). Instinct and survival: An exchange of letters between Einstein and Freud. *Canadian Medical Association Journal.* 163: 1178–1179.
Marris, P. (1996). *The Politics of Uncertainty*. London: Routledge.

Marx, K. (2007). *Economic and Philosophical Manuscripts of 1844*. London: Dover.
Mikulincer, M., & Shaver, P.R. (2007). *Attachment in Adulthood: Structure, Dynamics, and Change*. New York: Guilford Press.
Ostrom, E. (1990). *Governing the Commons: The Evolution of Institutions for Collective Action*. Cambridge, UK: Cambridge University Press.
Stuart-Smith, S. (2021). *The Well Gardened Mind*. London: Collins.
Tollemache, R. (2019). We have to talk about … climate change. In Hoggett, P. (Ed.), *Climate Psychology* (pp. 217–238). London: Wiley.
Tolstoy, L. (1873 [2010]). *War and Peace*. Trans A. Mandelker. Oxford: Oxford University Press.
Weintrobe, S. (2021). *Psychological Roots of the Climate Crisis*. London: Bloomsbury.
Wilkinson, R., & Pickett, K. (2009). *The Spirit Level*. London: Penguin.

Chapter 2

The world is on fire; why bother with psychotherapy?

James Angel

Climate change poses an existential threat to life as we know it. Already, in 2026, extreme weather events associated with rising global temperatures are causing severe suffering and loss across both human and non-human worlds (McSweeney & Tandon, 2024). Without fundamental political-economic change of the scale required to rapidly decarbonise society, the coming years and decades will see this destructive process escalate, with devastating consequences. Ecosystems will collapse; species will go extinct; poverty, inequality, ill health and mortality will soar – with the poor and people of colour the hardest hit (Parsons, 2023).

In the face of this inescapable yet unfathomable socio-ecological reality, what is the point of psychoanalytic psychotherapy? Does the unprecedented gravity of the climate crisis call into question the very meaning of the psychoanalytic encounter? Is psychotherapy rendered pointless, a redundant relic of a cooler planet? Or might psychoanalytic psychotherapy have something to contribute to the minimisation of suffering and the opening up of possibilities for hope in the context of the climate crisis? What, in short, is to become of psychoanalytic psychotherapy in a world that is on fire?

These are crucial questions for our profession. And, on a personal level, they are questions that frequently occupy my mind. Before training in attachment-based psychoanalytic psychotherapy at the Bowlby Centre, I worked in the climate change field – both as a campaigner within NGOs and as a researcher and teacher in academia. Looking back, perhaps my decision to leave this area of work and begin my psychotherapy training had something to do with the despair I felt. As fossil fuel usage continued and politics (both in my home country of the UK and elsewhere) shifted in even more reactionary directions, I felt defeated. In the face of this, I allowed myself to turn inwards, prioritising the working through of emotional wounds (primarily, of course, my own) over ecological wounds.

Yet I do not feel defeated enough to give up. Indeed, throughout and beyond my training, I have felt sure that my new calling of psychotherapy has something to offer in and to a climate-changing world. In what follows, I attempt to

DOI: 10.4324/9781003733942-2

distil some of my initial thoughts, feelings and instincts in this regard. This is very much a work in progress; a rough and partial sketching of scattered reflections intended as an invitation for conversation and disagreement.

To clarify, the chapter discusses the role of a specifically psychoanalytically informed psychotherapy within a climate-changing world. My assumption is not that psychoanalytic psychotherapy is necessarily the best-placed therapeutic modality for addressing climate change. Rather, my focus results simply from my own background as a psychoanalytic psychotherapist. This is the therapeutic orientation I am most familiar with and, in turn, best qualified to comment upon. Towards the end of the chapter, I do suggest that my own particular psychoanalytic lens of attachment-based psychoanalytic psychotherapy is in many ways helpful as a starting point for psychotherapeutic engagements with climate change. This, though, is of course not to say that other modalities do not have their own important and unique perspective to offer – I am sure that this is the case as well.

The chapter, then, is inevitably informed by my own individual therapeutic stance. Similarly, I have my own particular understanding of climate change. This, needless to say, also shapes the arguments made within the chapter – my take on the ways in which psychotherapy might productively contribute within and to a climate-changing world is very much informed by my personal analysis of climate change, its causes and its potential solutions. To be transparent on this point, then, my own view is that climate change is rooted in a capitalist mode of production that encourages limitless resource-extraction and consumption, that is wedded to sustained fossil fuel usage and that perpetuates huge inequalities (Klein, 2014; Malm, 2016). In addition, I see climate change as rooted in racist and colonial power dynamics, which value white lives above the lives of people of colour and that reproduce the exploitation of people and resources across countries of the Global South (Parsons, 2023). Accordingly, my opinion is that meaningful action on climate change means far more than individual lifestyle changes or technological shifts – as important as these forms of action might be. Rather, I see climate change as an alarm call signalling the need for very different ways of organising our societies, economies and environments. Unfortunately (or, perhaps, fortunately – depending on your point of view!) I do not have space within this chapter to elaborate on my views here in any further depth. I would say, though, that I would highly recommend Andreas Malm's (2016) *Fossil Capital: The Rise of Steam Power and the Roots of Global Warming*, Naomi Klein's (2014) *This Changes Everything: Capitalism vs The Climate* and Laurie Parsons' (2023) *Carbon Colonialism: How Rich Countries Export Climate Breakdown* as excellent resources for further reflection. Whether or not you agree with me on my analysis of climate change, my hope is that the paper might still prove interesting and productive for the question at hand of psychotherapy's relationship to the ecological crisis we find ourselves in.

The chapter is divided into three sections. In the first section, I suggest five possible ways in which psychoanalytic psychotherapy might contribute to a climate-changing world. In the second section, I offer some reflections on the implications of climate change for the politicisation of psychotherapy. Then, in a final concluding section, I give some brief thoughts on the importance of attachment within ongoing psychotherapeutic encounters with the question of climate breakdown.

What can psychoanalytic psychotherapy contribute to a climate-changing world?

In order for psychotherapy to play some kind of meaningful role in the context of climate change, it is incumbent upon psychotherapists to clarify what it is exactly about our profession that gives us something unique to say and do in this regard. Psychotherapists are not climate scientists; we do not have the know-how required to measure global temperature rises and their impacts. We are not engineers; most of us cannot build wind turbines. We are not campaigners (or not within our consulting rooms, anyway); we do not attempt to persuade our patients of particular political viewpoints. What is it, then, that we do that might have some use in addressing climate change? Below, I suggest five ways in which psychoanalytic psychotherapists might have something important and distinctive to offer.

1 *Psychotherapy can help us understand the psychological factors underpinning rising greenhouse gas emissions.*

It goes without saying that climate change raises many interesting and important questions pertaining to human psychology. We have known, for decades, that human behaviour is threatening the ecological conditions necessary for our survival. Yet, on the whole, our behaviour does not change – at least not in ways proportionate to the scale of the crisis at hand. This seems to say something quite profound about the nature of the psyche. Yet, all too often, discussions around the psychology of climate change seem a little simplistic: "human beings are just greedy and selfish"; "people are ignorant"; "most of us just don't seem to care"; "we're impulsive and hedonistic"; "we're just ostriches, inclined to bury our heads in the sand."

Psychoanalytic psychotherapy – as an expansive intellectual framework for understanding the psyche, grounded in rigorous clinical practice – can help offer some much-needed substance and nuance to the reductive psychological generalisations that abound in the public discourse on climate. From Freud onwards, analysts have been very interested in the ways in which, as human beings, we manage to get in our own way – this, if anything, is surely our area of expertise. Self-destructive behaviour, the repetition of harmful patterns, the masochistic pursuit of suffering – not a day goes by in the analyst's consulting room without meeting these contradictory aspects of the human experience.

Indeed, psychoanalysts have developed a host of concepts to help understand this – concepts that can also help us understand the psychic dynamics underpinning rising greenhouse gas emissions. To come to terms with the destructive impulses of his patients, Freud (1920) coined the concept of the death drive – a restless compulsion towards aggression that might be expressed towards both self and others. Meanwhile, his concepts of repression and repetition compulsion helped make sense of his patients' tendencies to act out counterproductive patterns, driven by unconscious forces (Freud, 1914, 1920). Contemporary analysts, while informed by Freud's original conceptual apparatus, have an array of other tools at their disposal to understand these paradoxical aspects of the human mind. Work on trauma and dissociation, for example, points towards the fragmentation of the psyche. Parts of the self are split off and disowned as "not-me," yet as much as their existence is denied, they remain powerful determinants of our everyday lives (Bromberg, 2003; Fisher, 2017). I know in myself, for example, I regularly swing between a highly consuming panicked urgency around the prospect of ecological collapse on the one hand, and a compartmentalised denial of the situation on the other. In this regard, my own internal relationship to climate change certainly speaks to something dissociative.

Psychoanalysis, then, offers multiple frameworks for understanding the ways in which much of human experience and behaviour is governed by complicated, irrational and paradoxical forces – forces, crucially, that are outside of conscious awareness. Today, contemporary psychoanalytic theory integrates and emphasises the role of the social world in the making and remaking of unconscious processes (Dalal, 2006) – from gender norms to consumer-focused advertising, what is repressed or dissociated cannot be understood without reference to the social, economic, and, I would suggest, ecological relations within which human minds are embedded.

Through the rigorous and slow-going exploration of our own minds and the minds of our patients, analysts have shaped an understanding of human experience that offers insights into our capacity for greed, ignorance, apathy, impulsivity, denial – and, at the same time, that sheds light on our capacity for care, empathy, protest and reflectiveness. We have, over the years, built a model of the mind that is able to hold together "good" and "bad," contradiction and ambivalence, conscious and unconscious, the social and the psychological. In grappling with the multiple complex psychosocial phenomena bound up in the production and reproduction of human-induced climate change, this model of the mind can surely be a powerful tool.

This, in fact, is demonstrated in Sally Weintrobe's important (2021) book, *Psychological Roots of the Climate Crisis: Neoliberal Exceptionalism and the Culture of Uncare*. Here, Weintrobe draws on her experience and knowledge as a practising psychoanalyst to explore the role of the psychic defence of "exceptionalism" within the climate crisis. The argument is that escalating greenhouse gas emissions must be understood as the product of neoliberalism, a dominant political-economic doctrine and policy framework based on free markets and the triumph of the sovereign individual over the collective interest. Neoliberalism, for Weintrobe, is psychologically grounded in exceptionalism, an omnipotent mindset of entitlement towards whatever one wants, irrespective of the ways in which reality might limit one's entitlements. Neoliberal "exceptionalism," in Weintrobe's account, blinds people to ecological limits in pursuit of personal enrichment and gain.

Weintrobe's account illustrates the productiveness of a nuanced psychosocial understanding of climate change that brings together the psychoanalytic and the political-economic lenses. Here, then, is a good example of the ways in which psychoanalytic thought can help inform our understandings of the origins of climate change – further work in this direction could also be illuminating.

2 *Psychotherapy can help us reflect on the impacts we have on ourselves and on others.*

To reiterate a point made above, one way of narrating climate change is as a story of humans struggling to contend with the destructive impact of their actions – on themselves, on others and on the ecological systems that life depends upon. In the above, I suggested that psychoanalytic psychotherapists have an array of tools to help diagnose why this struggle has come about. In addition, psychotherapy can help us navigate this struggle with greater skill and efficacy. Psychotherapy is surely one of the most effective tools humans have developed to help us reflect upon our behaviour and its impacts – both on self and others.

Most patients enter our consulting rooms because something about the ways that they are going about their lives is not working for them and/or their loved ones. Part of our job is to invite curiosity about habitual patterns such that, over time, our patients come to understand themselves, their minds and their behaviour in greater depth. One sign of an effective therapy is that patients begin to take more ownership and intentionality over their lives and their decision-making. Hopefully, processes of self-reflection and mourning, within the context of a secure therapeutic relationship, can help people put to rest self-destructive patterns and begin to make choices that better serve their interests.

My own conviction, though, is that our responsibility as therapists does not lie exclusively with our patients' personal interests alone. Rather, I am interested in the potential of psychotherapy to cultivate more ethically inclined and relationally adept subjects. When our patients behave in ways that harm those around them, we look to find compassionate ways to challenge this and to encourage our patients to understand, acknowledge and respect the subjectivities and needs of others. Peter Fonagy and Mary Target's (1997, 2002) work on mentalisation, for example, is a vital framework designed to help both therapists and patients with this. When my patients bring their interpersonal difficulties and conflicts to me, inspired by mentalisation-based approaches, I like to ask questions such as: "what do you imagine might have been going on in your partner's mind when they said that to you"; "when you sent that text message to your friend, how do you imagine they might have felt?"; "how do you think your sister might be experiencing your relationship with her at the moment?"

Climate change raises very interesting questions about human beings' capacity to reflect and mentalise, given that the consequences of fossil fuel-based consumption are distant in time and place. If I struggle to mentalise about the impacts I might have had on my friend in a disagreement we had yesterday, can I really be expected to mentalise about the impacts my decisions might have on a stranger in a far-away part of the world at some unknown, intangible future moment? This might seem like a tall order. However, our capacity for empathy and solidarity with strangers is evident across history and, indeed, daily life. Meanwhile, the capacity of psychotherapy to boost reflective capacity and mentalisation skills is evident across the countless patients we encounter and collaborate with. Perhaps there is the seed of something to work with here.

To clarify, my suggestion is not that the role of psychotherapists is to police and critique our patients' personal choices around, for example, eating meat or taking short-haul flights. As I note in the introduction to the paper, my own view is that conceptualising climate change as a question of individual consumer or lifestyle choices is a mistake. Rather, my view is that meaningful action on climate change requires a cultural shift towards a more collective, caring and publicly oriented way of inhabiting our world together. This, I would suggest, calls upon people to be more relationally attuned and skilled. Here, I think, is where psychotherapy might be able to help.

3 *Psychotherapy can help us gain an increased sense of agency – individually and collectively.*

One need not share my political analysis of climate change to acknowledge that powerful vested interests stand in the way of phasing out fossil fuels and decarbonising our societies, with many powerful industrial sectors – from energy firms to the meat industry, to the aviation sector – deeply invested in warding off change. As long as people feel powerless and hopeless, there is little chance of these vested interests being challenged. Therefore, achieving the kind of cultural shifts alluded to above in Point 2 will require people to claim a healthy sense of agency over their own lives – and over their relationship to the political and economic processes within which their lives are situated.

As the work of Judith Herman (1992, 1998) and many others has illustrated, experiences of trauma, abuse and neglect within our personal histories can undermine our sense of power, agency and control. Herman (1998, p. 98) argues that a central role of psychotherapy, especially with trauma survivors, "is to restore power and control to the survivor." Indeed, a vital task within psychotherapy more generally is to support individuals to gain a greater understanding of their own needs and limits and to become more confident, competent and forthcoming in advocating for these.

In turn, psychotherapy can facilitate a vital recovery process of agency being reclaimed. Changes in this direction in one's personal life and immediate relationships can often translate into an increased appetite to be seen and heard at the collective level. For as we begin to grasp that we deserve to be treated with decency and dignity by those around us, the extent to which these basic rights are often denied socially and politically becomes all the more clear – as does our sense of entitlement and self-belief to do something about this. Supporting our patients to claim political agency in this way, then, becomes an additional way in which psychotherapists can contribute to endeavours to address climate change.

My view is that encouraging our clients to explore their own sense of agency, fairness, justice and power is very much part of a psychotherapist's remit – and that this need not be confined to the meaning of these ideas within the individual realm alone. This, of course, has to be done sensitively and non-didactically, led by the patient's own curiosities rather than the therapist imposing their own agenda. Yet, sometimes, I notice in my own clinical practice a caution to move beyond the individual level within sessions, out of an anxiety around being too directive or leading – I imagine that I am not alone in this experience. The relationship between the individual and the collective, and how this can be navigated within the clinical space, is a complex topic. Given the existential stakes of the climate crisis, it is more important than ever that therapists engage with this complex question with nuance and respect – I return to this topic later on in the chapter.[1]

4 *Psychotherapy can help us mourn the losses of climate breakdown.*

As the destructive consequences of climate change and ecological collapse unravel, the concept of ecological grief is increasingly discussed (Cassidy-Williams & Cunsolo, 2018). Indeed, sadness, despair and anxiety around the multiple losses bound up in climate breakdown – including the loss of beloved places, species and people – are increasingly entering psychotherapists' consulting rooms. Therefore, psychotherapists have an important role to play in addressing these losses and subsequent experiences of grief at both an individual and collective level (Vallianatos, 2020). In other words, psychotherapy can help us mourn the losses of climate breakdown.

The idea of mourning is, of course, at the centre of many therapeutic practices, none more so, perhaps, than attachment-based psychoanalytic psychotherapy. John Bowlby wrote extensively about the role of mourning within the formation of distress and well-being. Drawing on his observations of children separated from their parents alongside those suffering bereavements, Bowlby (1980, 1988) identified four phases in the mourning process. Mourning, for Bowlby, begins with a period of numbness within which emotional capacities shut down and the reality of the loss is not faced. After this numbness comes a lengthy period of "yearning and searching for the lost figure" (Bowlby, 1980, p. 85). This consists of an intense longing for the lost object and a continual working over of the events leading up to the loss in the mind of the mourner, underpinned by the wish that things might still turn out differently. This aspect of the mourning process, Bowlby writes, often sees the emergence of powerful feelings of anger directed at all of those who might be responsible for the loss, including the lost object themself. In phase three of Bowlby's mourning process comes disorganisation and despair: the mourner's life has been thrown into turmoil, and a period of depressive hopelessness and aimlessness often ensues. It is, for Bowlby, only after these three phases of mourning have been worked through that the final phase of "reorganisation" can begin. At this point, the mourner can finally come to accept the reality of the loss and, in turn, the possibility that a new life with alternative forms of meaning and vitality can be found.

It seems to me that Bowlby's account of mourning might be helpful in addressing the losses entailed within climate breakdown. When it comes to this question, perhaps many of us are stuck in the first three phases of mourning. We might find ourselves permanently numb and shut down to the devastating reality of what has already been irredeemably lost to ecological crisis, alongside the magnitude of the losses to come. We might find ourselves furious with governments and fossil fuel companies, wishing that if we can just protest enough, then lost species, livelihoods and lives will be regained. We might find ourselves utterly lost, hopeless and passive, frozen by the scale of the losses at stake. All of these responses are understandable and important. In particular, I think that the anger typical of phase two is absolutely legitimate and vital as a fuel for social movements endeavouring to hold the powerful accountable for the crisis we find ourselves in. Yet, crucially, Bowlby's point is that the first three phases of mourning must be worked through in order to then find a response to loss that is truly constructive and creative. Only when we meaningfully and genuinely accept what we have already lost and continue to lose to the climate crisis, can we – as individuals and societies – truly appreciate the reality of the situation and begin to reorganise our lives accordingly.

As such, a psychotherapy informed by Bowlby's account of mourning can help us feel and process our ecological grief and, in the process, move towards a psychic position more amenable to change, action and repair. This, in turn, might help us go on to reduce the scale of future climate loss and harm.

5 *Psychotherapy can help us navigate the everyday suffering of climate change.*

Representations of ecological crisis in mainstream culture often depict dramatic apocalyptic scenarios of sudden collapse. When understood this way, climate change can seem like some kind of distant one-off event: a sudden "tipping point" sometime in the future from which there can be no return. This narrative, however, can be quite misleading (Bellamy & Hulme, 2011). Rather than being a dramatic one-time rupture issuing in sudden social breakdown, climate change is better understood as a gradual, complicated and non-linear process of social and ecological transformation – a process that will be experienced and felt within our everyday lives through multiple phenomena, from rising food and energy prices to increased forced migration, from more frequent extreme weather events to intensified conflict over resources.

Climate change, then, can in one sense be understood as a process very much bound up with our everyday suffering. The extent of the everyday suffering at stake depends upon the effectiveness of the climate mitigation and adaptation measures that our societies are able to deploy. Yet even in the most optimistic of future scenarios in this regard, a substantial rise in global temperatures already seems inevitable, meaning that significant everyday suffering will ensue – indeed, this is happening already (McSweeney & Tandon, 2024). This suffering, of course, will primarily need to be addressed politically through measures that make a material difference, such as better healthcare services, more stringent regulation of the economy, more welcoming and inclusive migration policies and mass investment in low-carbon infrastructure (Klein, 2014). Yet, alongside this, there will need to be some acknowledgement that increased material poverty and hardship will inevitably mean a deepening of emotional and psychological suffering. Life is likely to become more uncertain, unpredictable and anxious. Providing a secure enough environment for ourselves and our children is likely to become more difficult. The trauma inflicted by extreme weather events will become more widespread. As is argued above, people will increasingly be forced to contend with climate-related losses and grief.

In sum, if climate change is to increase human suffering in these ways and others, there will be all the more need for psychotherapy as a space in which our psychological suffering can be understood, contained and reduced. The risk, of course, is that with climate change likely to accentuate pre-existing social and economic inequalities, for example, those shaped through social relations of race and class, access to psychotherapy becomes even more polarised than it is already. Getting real on this risk, I believe, requires a politically engaged psychotherapy. What I mean by this is discussed in more detail in the next section of the chapter.

Climate change and the politics of psychotherapy

Thus far, this chapter has sketched some initial thoughts on various ways in which psychotherapy might make a productive contribution within and to a climate-changing world. I have suggested that the model of the mind developed by psychoanalytic psychotherapy can help us understand the psychological factors underpinning human-induced climate change. I have suggested that, in helping people to reflect upon the impact of their behaviour and to develop an enhanced sense of agency, psychotherapy can make a meaningful contribution to efforts to tackle climate change. And I have suggested that by helping people to mourn the losses of climate change and to navigate the suffering bound up with climate change more broadly, psychotherapy has a role to play in supporting individuals and societies to work through and adjust to the distressing realities of a warming world.

Implicit in these suggestions, I believe, is a psychotherapeutic practice and profession that is unafraid to be politically engaged. This idea is likely to arouse multiple different responses and associations – I anticipate that some colleagues may well believe that politics should be left out of the consulting room and the psychotherapeutic world more broadly. At this point, it is worth clarifying that I am not advocating for our consulting rooms becoming lecture theatres or campaign battlegrounds. We must, of course, not start to lecture or persuade our patients of particular political perspectives – this could lead to highly unethical and counter-therapeutic exchanges. I want to conclude the paper, then, by saying a little more about what I have in mind by a politicised psychotherapy in the context of the climate crisis.

First, coming back to an earlier point, I think that the exacerbation of poverty and inequality that climate change likely implies makes conversations around the accessibility of psychotherapy all the more important. The injustice of climate change is that the poorest tend to be the least responsible for greenhouse gas emissions and, at the same time, the hardest hit by the impacts of climate breakdown. If psychotherapy, as per my point above, is to play a role in helping those on the frontline of climate change to navigate their psychological suffering, then much will have to change around the financial accessibility of our practices. On the one hand, this means transforming state provision of free psychotherapy through public healthcare systems (in the UK context, this means the NHS) with vastly increased funding and substantially increased provision of long-term psychotherapeutic treatment. In addition, psychotherapists in private practice might consider taking on additional low-fee work and, as many of us do already, implementing sliding scale fee structures.

One innovative idea already being implemented by the *Climate Psychology Alliance* is a therapeutic support service that offers three free sessions to those dealing with distress related to the climate and ecological crisis (see Climate Psychology Alliance, 2024). Many of those benefiting from this support are climate activists grappling with the grief and despair that accompanies sustained climate campaigning work, not to mention the stress and anxiety of court cases and, in recent cases, prison sentences for those who have taken direct action and been arrested with groups such as *Extinction Rebellion* and *Just Stop Oil*. To my mind, the Climate Psychology Alliance initiative is valuable not only because of its efforts to make psychotherapy more financially accessible. Further, the value of this endeavour is in its explicit intention to provide therapeutic support to those pursuing meaningful social change with the goal of minimising the impacts of climate breakdown. Here, we see psychotherapists really aligning themselves with practical efforts to address climate change in a way that foregrounds the thing we do best: practising psychotherapy!

Perhaps less direct, but equally important, is that psychotherapists actively campaign for policy changes that would help expand access to the meaningful long-term psychotherapy required to address ecological grief and climate-induced emotional distress. If, as I have suggested, climate change likely means worsened psychological suffering for many, the work of our professional bodies, training organisations and campaign groups in advocating for fundamental change within the mental health system becomes all the more important. Under-funded services, long waiting lists and an emphasis on short-term manualised therapeutic approaches do not constitute a system fit for purpose – a reality that will only become more stark as the psychological impacts of climate change deepen. Climate change, then, adds extra impetus to pre-existing campaigns for fundamental change within state-funded mental health provision. Yet our public campaigning and advocacy work need not end there: given that climate change is a significant threat to our individual and collective well-being, then surely mental health professionals have the legitimacy and responsibility to speak out on this issue more broadly. This, indeed, is the position of groups such as *XR Psychologists*, a group comprised of psychotherapists, psychologists and psychiatrists. In their words:

> The climate and ecological emergency that faces us all…will have massive implications for the well-being, health and effectiveness of all humans and all other life on the planet. Many psychologists see the impact that this crisis will have on their areas of work, their families, communities, and the wider world. There is no simple psychological fix. But as national and international governments fail to push forward with meaningful, timely, concrete actions – psychologists are taking a stand and joining this movement for change.
>
> (XR Psychologists, 2024)

Part of the power of initiatives like the Climate Psychology Alliance and XR Psychologists is that they move beyond the level of the individual and towards the collective sphere. They do this, first, by coalescing clinicians to reflect and act together, rather than alone within the silos of our own practices. In addition, they help clinicians to broaden their outlook from the lives of their individual patients and towards the broader social context within which the lives of our patients – and, of course, our own lives, both personally and professionally – are embedded. To me, this collective outlook is essential to a meaningful psychotherapeutic response to climate change. As alluded to previously, my own view is that climate change is, in part, a product of an overly individualistic culture. As is argued by Weintrobe (2021), neoliberalism encourages us to maximise our own personal gain, irrespective of the impact on others, or the impacts on the ecological systems that sustain us. What's more, neoliberal solutions to climate change have largely been premised upon individual lifestyle changes and "green" consumerism – solutions that have failed spectacularly to meaningfully curb fossil fuel usage or greenhouse gas emissions (Klein, 2014).

In contrast, effective climate action demands collective thinking and action, from organised civil society pressure through to strong government planning, coordination and leadership. Critics of psychotherapy, however, have long argued that therapy reinforces the individualism of neoliberal culture, encouraging patients to locate the source of their distress in the particularities of their own minds, histories and families, rather than the social and political worlds they inhabit (Gergen, 2009; Frosh, 2010). As relational psychoanalytic psychotherapists, we know that the individual psyche and the social world cannot be divorced in this way (Dalal, 2006). It is, then, incumbent upon us to put theory into practice and to be more oriented towards the collective sphere. This is a question of the kinds of lenses, questions and interventions we deploy with our patients. And it is a question of how, as a profession, we engage with the world around us.

That said, as well as being more outward-looking and collectively minded, I also believe that psychotherapists need to stay humble and mindful of our own omnipotent fantasies. Psychotherapy, as I have suggested here, does have the potential to make a meaningful contribution within and to a climate-changing world. Yet we need to be very clear that the contribution that we can make as psychotherapists is, perhaps, quite limited. Therapy alone is not going to halt climate change – as I have been stressing throughout the chapter, this is a highly complex socio-ecological problem rooted in deeply entrenched political-economic processes and systems. The idea, which I have sometimes heard expressed, that the answer to climate change is government leaders and fossil fuel executives going to therapy, strikes me as a very dangerous delusion. There is a risk that, in our well-intentioned desire to offer something on climate change, psychotherapists fall down a well-worn track of reducing social problems to psychological problems. Farhad Dalal (2006), in his analysis of psychotherapy's engagement with racism, has written compellingly on this. Dalal is critical of a tendency within the psychoanalytic community to assume that all distress – both on the level of the individual psyche and on the level of the external social world – is, at base, rooted in psychological factors. He writes:

> Interestingly, much of psychoanalytic thinking has been reluctant to allow the external social world to play a causal role in the structuring of internal distress, and, instead, is more likely to explain difficulties in the external social world as due to, and driven by, distress in the internal world … My intention is not to dismiss internalist psychoanalytic explanations of social phenomena out of hand, but rather to point out their limitations. I find these ways of thinking useful and essential to my clinical work; however,

> when they are put forward as the only explanations and pose as being complete, then they lose value and become dangerously reductive – in which case, they do not so much explain as explain away.
>
> (pp. 138–139)

Conclusion: Attachment in a climate-changing world

It is in the spirit of Dalal's words that I intend this paper to be understood. I have argued that psychotherapy can find a place in a climate-changing world by attending to some of the "internal" dimensions of the climate crisis. Yet, in doing so, it is crucial that we do not assume that psychotherapy and what Dalal calls "internalist" explanations have all the answers to a crisis that is very much rooted in the "external" social world. Attachment theory, for Dalal (2006), is one approach that has the potential to avoid the reductive "internalist" lens that many other psychoanalytic modalities fall into. This is because, following Bowlby, Attachment Theory has traditionally distinguished itself from psychoanalytic orthodoxy by emphasising the importance of "external" real-world events and relationships in the formation of well-being and distress, rather than assuming the primacy of internal fantasies and drives (Holmes, 1993).

Building on Dalal's thinking here, I want to conclude with some brief reflections on the place of a specifically attachment-based psychoanalytic psychotherapy within a climate-changing world. As mentioned in the introduction, writing as an attachment-based psychoanalytic psychotherapist, I offer no pretence of an objective or impartial take here. That said, my impression is that Attachment Theory and attachment-informed clinical practice do have much to offer the wider psychotherapeutic world when it comes to the question of our engagement with climate breakdown.

First, earlier in the chapter, I argued that helping people mentalise and mourn constitutes two important ways in which psychotherapy might be useful in the context of climate change. Mentalisation and mourning are, of course, two of the fundamental pillars of attachment-based clinical practice. In this sense, then, Attachment Theory makes for an important theoretical foundation for some of the contributions I have advocated for in this paper. Indeed, my instinct is that some of the additional core concepts of Attachment Theory will likely prove useful in further psychotherapeutic engagement with climate change. Perhaps one way of understanding our psychological relationship to climate change is through the threat of the loss of the secure base of a hospitable planet. Might immobilising terrors of apocalyptic collapse reflect something of a preoccupied state of mind? Could we see attempts to deny or minimise the problem of human-induced climate change as an expression of a somewhat avoidant state? Is there not a disorganised quality to attempts to plough on with continued fossil fuel extraction, not to mention fantastical visions of geoengineering (from mirrors in space through to fertilising the ocean)? In tentatively raising these questions, I am wondering whether

Attachment Theory might have something important to add to the first contribution suggested above, namely, attempts to use the tools of psychoanalytic psychotherapy to help us understand the psychological factors driving the climate crisis.

Moreover, as stressed by Dalal, Attachment Theory's refusal to reduce everything to the internal or psychological is an important starting point for a more politically engaged psychotherapy. Indeed, Bowlby himself saw a need for the psychotherapeutic community to be politically involved. Himself a committed socialist, Bowlby was unafraid to advocate for the policy changes he believed that his research on attachment justified – from hospital reform through to state funding for families of troubled children (Holmes, 1993; Mayhew, 2006). Indeed, Bowlby grew frustrated with his colleagues' lack of political commitment. He railed against the British Psychoanalytic Society's refusal to engage with the establishment of the National Health Service post-Second World War, commenting:

> We find ourselves in a rapidly changing world and yet, as a society, we have done nothing, I repeat nothing, to meet these changes, to influence them or to adapt to them. That is not the reaction of a living organism but a moribund one. If our society died of inertia it would only have met the fate it had invited.
>
> (cited in Holmes, 1993, p. 24)

We might wonder if Bowlby would find similarly candid words in relation to the psychotherapeutic world's engagement with the "rapidly changing world" we see now, in the context of climate breakdown. In a world on fire, there is surely something to learn from Bowlby's capacity to challenge the "inertia" he observed within the psychotherapeutic community – and his willingness to speak out in favour of our profession's capacity to meet change, to influence and to adapt.

***

Note

1 I want to credit Sue Berger for her role in inspiring my reflections in Point 3. My discussion with her around the politics of psychotherapy and the role of psychotherapy in supporting patients to find their own sense of agency helped reflect on my own position around this important question.

References

Bellamy, R., & M. Hulme (2011). Beyond the tipping point: Understanding perceptions of abrupt climate change and their implications. *Weather, Climate and Society.* 3: 48–60.

Bowlby, J. (1980). *Attachment and Loss, Volume 3, Loss: Sadness and Depression.* London: Hogarth Press.

Bowlby, J. (1988). *A Secure Base: Clinical Applications of Attachment Theory*. London: Routledge.

Bromberg, P.M. (2003). One need not be a house to be haunted: On enactment, dissociation, and the dread of "not-me" – A case study. *Psychoanalytic Dialogues.* 13(5): 689–709.

Cassidy-Williams, M.K., & Cunsolo, A. (2018). Ecological grief and the intersection of climate change, identity, and well-being. *Environmental Sociology.* 4(1): 1–8.

Climate Psychology Alliance. (2024). One-to-One climate aware therapeutic support. Available at: https://www.climatepsychologyalliance.org/index.php/component/content/article/one-to-one-therapeutic-support?catid=13&Itemid=101 (accessed 19 November 2024).

Dalal F. (2006). Racism: processes of detachment, dehumanization, and hatred. *Psychoanalytic Quarterly.* 75(1): 131–161.

Fisher, J. (2017). *Healing the Fragmented Selves of Trauma Survivors: Overcoming Internal Self-Alienation*. New York: Norton & Company.

Fonagy, P., & Target, M. (1997). Attachment and reflective function: Their role in self-organization. *Development and Psychopathology.* 9(4): 679–700.

Fonagy, P., & Target, M. (2002). Mentalization and the changing aims of psychoanalytic treatment. *International Journal of Psychoanalysis.* 83(6): 1051–1068.

Freud, S. (1914). *Remembering, Repeating, and Working-Through*. In *The Standard Edition of the Complete Psychological Works of Sigmund Freud*, edited by James Strachey, vol. 12, pp. 145–156. London: Hogarth Press.

Freud, S. (1920). *Beyond the Pleasure Principle*. In *The Standard Edition of the Complete Psychological Works of Sigmund Freud*, edited by James Strachey, vol. 18, pp. 7–64. London: Hogarth.

Frosh, S. (2010). *Psychoanalysis Outside the Clinic: Interventions in Psychosocial Studies*. Basingstoke: Palgrave Macmillan.

Gergen, K. (2009). *Relational Being: Beyond Self and Community*. Oxford: Oxford University Press.

Herman, J. (1992). *Trauma and Recovery: The Aftermath of Violence – From Domestic Abuse to Political Terror*. New York: Basic Books.

Herman, J. (1998). Recovery from psychological trauma. *Psychiatry and Clinical Neurosciences.* 52: 98–103.

Holmes, J. (1993). *John Bowlby and Attachment Theory*. London: Routledge

Klein, N. (2014). *This Changes Everything: Capitalism vs The Climate*. New York: Simon & Schuster.

Malm, A. (2016). *Fossil Capital: The Rise of Steam Power and the Roots of Global Warming*. London: Verso.

Mayhew, B. (2006). Between love and aggression: The politics of John Bowlby. *History of the Human Sciences.* 19(4): 19–35.

McSweeney, R. & Tandon, A. (2024). "Mapped: How Climate Change Affects Extreme Weather Around the World." *Carbon Brief.* Available at: https://interactive.carbonbrief.org/attribution-studies/index.html (accessed 19 November 2024).

Parsons, L. (2023). *Carbon Colonialism: How Rich Countries Export Climate Breakdown*. Manchester: University of Manchester Press.

Vallianatos, H. (2020). Healing ecological grief: Ecopsychology as a means of emotional support in the face of climate change. *Psychology and the Environment.* 24(4): 341–359.

Weintrobe, S. (2021). *Psychological Roots of the Climate Crisis: Neoliberal Exceptionalism and the Culture of Uncare*. London: Bloomsbury.

XR Psychologists. (2024). XR Psychologists declaration. Available at: https://xrpsychologists.co.uk/ (accessed 19 November 2024).

Chapter 3

Nature and belonging

Attachment and separation from the Great Mother

Roger Duncan

In this chapter I am intending to take you on a bit of a journey, and it's something of an imaginal journey. In other words, it's a journey where I use images that have an emotional impact to provide the scaffold for a way of knowing, and I hope this will become clear as we go. I will start with a little bit about my own relationship with attachment.

I'm an adoptee, my father was a doctor, a GP in the UK, and my mother was a housewife, and I was adopted from Canada. My birth mother was Canadian, and I eventually met her when I was 40. My adoption was relatively straightforward in one sense; I never experienced any kind of abuse or neglect, unlike a lot of the young people I work with as a psychotherapist. I merely changed from one mother to another when I was about 6 weeks old. But it has had a profound systemic impact on my life, and it has taken me a long time to make sense of this and how it has influenced my thoughts and emotions throughout my life.

My father was sent to a preparatory school and then a boarding public school as a young man during the Second World War. I, too, was sent to boarding school from the age of 8 years old up until the age of 18. I had the whole boarding school experience, but again I did not experience any abuse there, although some of my friends were not so lucky. But this experience of separation had a big impact on my life. There is now a lot of research on the impact of boarding school, and making sense of attachment has been a large part of my interest and training to become a psychotherapist. And a big part of this journey for me has been the impulse to connect with nature as a secure base.

Essentially, I've come to understand that I was looking for my mother in nature, and reaching the age of 65, I have also learnt that if you spend much of your life looking for something, you eventually find it. I have found in my research and experience that nature offers a deep sense of attachment and belonging, and this sense of connection has often been personified in traditional cultures as a reconnection with the Great Mother, an idea I will return to later in this talk.

What I find interesting is when you start to look for things, when you start to look for them with intensity, interesting things turn up, and this has also been part of my journey.

DOI: 10.4324/9781003733942-3

For this audience, I don't need to tell you much about attachment, because many of you have a background and training in psychotherapy. But part of what I have been exploring in my new book (*Nature and Belonging*, in print) is the idea of moving from understanding attachment patterns as part of psychological trauma, to seeing attachment as longing for belonging, a deep need that we all have, to connect.

As you well know, much of our current understanding and ideas about attachment have come from the work of John Bowlby (Bowlby, 1979) and Mary Ainsworth (Ainsworth, 1967), who developed a kind of academic and medical response to this deep need for us to connect. This approach has now become part of our understanding of psychotherapy and our approach to social care. And I suppose the reason I frame it that way is that somehow older patterns and systems of belonging got lost along the way, and that is what I hope to explore in this talk.

However, these older systems of belonging that we can trace back to pre-industrial times are not completely lost; we are still in contact with them, they are right here, and they are very deep.

Another model of attachment that some of you may have come across is the dynamic maturational model of attachment developed by Pat Crittenden, who is a psychologist and systemic psychotherapist (Crittenden, 2008). She explores an understanding of attachment as a sense-making cycle, where accurate sense-making is a combination of accurate thinking and feeling. She describes how disrupted attachment results in either over-reliance on thinking at the expense of understanding our feelings, or an over-reliance on feeling with little capacity to think clearly. Using this model of attachment in psychotherapy directs us to help balance a clear experience of both our emotions and our thinking. It's a very helpful model to work with.

However, more recently I've come across a very interesting and inspired model of attachment in the book *The Evolved Nest* by Darcia Narvaez (Narvaez & Bradshaw, 2023). She explores different parenting and attachment patterns in animals and birds. Her work is also deeply informed by traditional healing practices that have supported healthy human development for 30,000 generations prior to industrialisation. She describes nine different practices that have been traditionally part of creating healthy human cultural attachment patterns.

These include creating a soothing perinatal social environment for mothers, prolonged breastfeeding, positive and caring touch from the mother in early life, including holding and rocking, and creating a culture where children are unconditionally welcomed into a supportive community of adults. She also describes the essential need for childhood play in mixed-age groups, as well as contact with nature and regular healing practices within a community (Narvaez, 2023).

Narvaez points out in this well-researched book that healthy attachment is not a psychological issue for individuals, but is a cultural issue based on connection to a healthy family and community. She describes how attachment issues are not located within individuals, but the problems are located in the culture, and this requires structural change.

The task of creating a culture that provides supportive environments is challenging because we must change the story, the narrative, of what is required for secure human attachment, and this requires an acknowledgement of the limits of Western thinking and culture around attachment and nature connection. Seen in this way, many of the attachment problems we see have come about in the context of Western culture with its tendency towards categorisation at the expense of relationships, where the wild otherness of nature and complex mental health presentations are reduced into a mere taxonomy, just labelling. We can see this in mental health work, where there is a tendency to attempt to fit people into a diagnostic category, leading to medication or a treatment pathway, a process that, as we know, can have its limitations. But we also see this in biology, where there is a strong tendency towards naming and explaining all that happens in nature. One of my formative experiences of working in the woodlands at Ruskin Mill College for over ten years with complex needs adolescents was that I built a relationship with the woodlands which was beyond merely identifying the plants. I knew the Latin name of the bluebells, *Hyacinthoides non scripta*, but I had never before had the experience of seeing bluebells grow and die away each spring for over ten years. Learning the names of the plants and trees is a completely different experience from actually living with and working with them. This difference between naming and experiencing is something I have been exploring for a long time. Gregory Bateson calls this the "epistemological error," the difference between how humans think and how nature actually works, which eventually leads to an error in how we think about things, so we don't get to see how nature really works (Bateson, 1979).

John Bowlby was a great admirer of Darwin and even ended up writing a biography of the great biologist (Bowlby, 1990). But post-Darwinist ideas have been reduced to the construction of social narratives, social Darwinism, where the narratives about evolution tend to be very linear, first-order and teleological. A modernist world view that lacks any real curiosity, and an implicit belief in the unconditionally positive value of teleological process and technocracy. The privileging of technology over any form of relationship that is not embedded in the most advanced technology, an advancement that is actually not sustainable.

This belief in technocracy is underpinned by a deeper, implicit belief in the centrality of unregulated consumerism and hyper individualism, ideas that were brought to America in the late 1940s by Sigmund Freud's nephew, Edward Bernays (Tye, 1998). With his uncle's ideas in his head, Bernays realised that by understanding the mechanisms and motives of the group mind it was possible to control the masses according to our will without them knowing. He realised that he could use people's unconscious urges to make them buy things they did not really need and could be manipulated into a marketing strategy, an idea he then sold to early marketing companies in America, which later spread throughout the world. Bernays famously opened the predominantly male market of cigarettes to women, which he branded as "torches of freedom," by placing famous female actresses smoking in popular movies of the time. This work was essentially a misuse of psychoanalysis and a transgression of ethical boundaries to create the modern consumer culture. Advertising is now an unchecked ecological safeguarding issue, which no one is flagging, encouraging people to buy things they don't need and stripping resources from the earth. We end up with the reality of climate change, which has flushed out the fact that current Western epistemology, the Western way of understanding the world, is now a dangerously outdated operating system. It just doesn't work anymore, but the problem is, we don't have a better system. Western culture is trapped in an epistemology where both mind and nature are considered to be just complicated physical mechanisms that can be explained and fixed by human thinking.

One reason we don't hear much about the climate crisis on the news or in the newspapers is that there is a vested interest in maintaining business as usual, even while the system is failing. However, this is being highlighted by thinkers, such as Daniel Schmachtenberger (Schmachtenberger, 2020) and his understanding of the meta-crisis, a combination of compound systemic issues that include not only social crises, but issues like microplastics in the ocean, overfishing, destruction of the environment, and forever chemicals in the rain, preventing human fertility. When we explore this, it is clear that the situation is very bad; this global crisis is inevitable, unpredictable, and irreversible, and will become more so.

But there are some ways to think about this more positively, and here I can direct you to the work of Benjamin Costello, who has combined an understanding of the meta-crisis with the Kübler-Ross's stages of grieving (www.newworldtogether.org). He starts with the first stage, denial of our current state, and the justification of maintaining the business-as-usual approach to our work. The next stage is the shock we experience with news of the latest climate disaster, a flood or fire. This can lead to emotional overwhelm and an urgent need to find a solution or the need to blame somebody. The next stage leads us into an existential crisis where nothing makes sense anymore, a difficult place to be, where our old sense- making strategies no longer work for us. We then move into the next stage, of acceptance that things are changing in ways that we were not expecting, an awareness that the core of the problem is within ourselves and how we approach the problem. The final stage described by Costello is an embodiment of the problem and the taking part in the challenging social and psychological process of rehabilitation.

In my role as a psychotherapist, this is helpful to me and points towards a future path for psychotherapy in the meta-crisis; it helps me think about the role of psychotherapy within a changing global culture. This is part of the work that I have been involved in for the last 40 years, which has now got a name, Systemic Ecopsychotherapy. This work for me started with connecting young people to nature, which has also been part of my own journey in finding healing for myself. This has been a process of recognising that spending time in nature and working with the body is profoundly therapeutic work. During this time, I have witnessed young people with extreme difficulties and challenging behaviour change quite radically, through engaging with their bodies and working in nature. This work, which includes wilderness therapy, involves long exposure to nature away from Western culture for up to ten days at a time, taking all the food we need. I have seen young people transformed in this process after only three or four days.

Systemic Ecopsychotherapy supports mental health, essentially through contact with nature, and the evidence shows that this involves three elements, which are:

- Being outside in nature
- Engaging with a social group
- And working in a meaningful context.

This is not surprising, as this is what Indigenous cultures have been doing for thousands of years and the last 30,000 generations. But to access this experience within Western culture, we must dig quite deep; it's actually quite difficult. These difficulties are, however, not clinical but structural, which makes it difficult to create programmes, particularly for adolescents, because they are busy with exams in school, which are usually considered more important.

Part of the work I've been doing is trying to develop skills, practices, and interventions for adolescents. My research has focused on three Ecopsychotherapy practices, each of which can lead to the other two.

These three practices are "Healing developmental and transgenerational trauma," "Re-storying our current view of the world" and "Reconnecting with nature."

Healing trauma

The work of healing trauma starts with the body, and this is some of my main work in CAMHS (Child and Adolescent Mental Health Services) with adolescents, as most of the mental health issues and behavioural problems that we see in CAMHS can quickly be traced back to trauma and transgenerational trauma. As you may know, our understanding of trauma has now moved from even 10 or 20 years ago. And if we look at the meta-crisis we are facing, we might say that much of the behaviour we see worldwide, the destruction of the ecosystems, is the result of a culture driven by unprocessed trauma. As we know from working with adolescents who are self-harming, without a process of self-reflection, such as psychotherapy, unprocessed trauma can lead to destructive relationships with the self and the world. It seems likely that this is what is happening in the world: the legacy of unprocessed transgenerational trauma is fuelling our unhealthy and destructive relationships with nature.

Galina Angarova, who was formally the United Nations spokesperson for Indigenous people, describes how she grew up as a part of the Buryat Indigenous peoples near Lake Baikal in Russia (Angarova, 2020). She lived through the transition from living within an Indigenous village, isolated from modernist culture and witnessed the transition to a money-based consumerist culture and saw her tribe make this transition. She relates that her tribe is now highly represented in the professional classes, the doctors and the lawyers in that region. She believes this is a trauma response, like Stockholm syndrome, where the victims of trauma take on the characteristics of the perpetrators. And she describes her journey of reclaiming her Indigenous connection after having been Western educated and trained in America. She went back to her tribal village where she grew up, and realised she still had a deep connection with the land, which she had to relearn through overcoming her Westernised worldview.

This is a common story I have heard in the biographies of many Indigenous people, who have to learn to shed their Western worldviews before they can reclaim their Indigenous culture.

American trauma therapist Resmaa Menakem has made a study of colonialism as a trauma state, how the roots of colonialism can be traced back to transgenerational trauma in Europe before the settlers even arrived in the Americas. This trauma occurred in the Middle Ages and Dark Ages, which included the European witch trials and the mass killing of women, and how this impacted the culture (see Menakem, quoted in Tippett, 2020). In a podcast interview with Resmaa Menakem, he said famously, "trauma in a person decontextualised over time looks like personality." We see this in politics, where mostly male, right-wing politicians often present as larger-than-life personalities. Menakem goes on to say that "trauma in the family, decontextualised over time, looks like family traits"; *a* familiar idea among family therapists and something I work with in the family therapy clinic on a weekly basis. But he also adds, "*trauma in a people decontextualised over time like culture.*" So just as an individual can be traumatised, cultures also carry trauma, and although we have structures, such as psychotherapy, to heal individuals we don't have structures for healing cultures currently.

If you consider yourself a non-Indigenous person, and are not clear if you have experienced transgenerational trauma within your culture, you might want to ask yourself, why are you not speaking your tribal language anymore? And when was this lost? Much of this loss of tribal European Indigenous identity and language dates back to Roman times, where there was a deliberate colonial intent to break the Celtic tribes of Britain. Where I live in the Southwest of the UK was the land of the Duboni tribes, and close by was one of the last stands of the tribal group in Britain, at a place still called "*The*

Slaughters" on the River Wye, where many of the last Celtic tribes were killed or captured by the Romans and taken back to Rome.

To move beyond a colonial mindset and narrative within Western culture, we need to make what Maldonado-Torres describes as the decolonial turn (Maldonado-Torres, 2004), to get to a position where we move beyond the assumption that the world cannot get by without the thinking of a Western subject and learn to trust that non-Western and Indigenous cultures have ways of knowing that are not dependent on research and evidence, so that even research and methods of study become decolonised. Traditional Indigenous knowledge systems were brutally suppressed and eradicated through the colonial system of boarding schools, which were used as a brutal method of colonial suppression. The legacy of this practice takes us to our second practice of trying to find a different story of the world.

Re-storying the world

Re-storying the world requires us to take seriously other different and challenging epistemologies of nature and mind, being really curious concerning different ways of thinking about and seeing the world. This requires the difficult shift away from our familiar ways of knowing and to embrace a stance of not believing what we believe (Kripal, 2020), for example, moving our view or gaze of the world from a Western perspective of Newtonian space and time as the basis of reality to seeing the world from a more ancient perspective. Two examples I explore here are the Aboriginal dream time and the Buddhist and Hindu image of Indra's net. To understand this change we need to approach these epistemologies, not as subsets of modernist thinking. They are understood best not as belief systems, because this already positions them not as versions of reality in the way space-time is considered to be, but just archaic forms of understanding the world. What I am suggesting here is to think of these as different lenses through which we can gaze at the world, and through which a different world then appears.

Both the dream time and Indra's net offer a relational way of seeing the word, where the primary matrix of reality is not objects, things, space and distance, but the relationship between things, and the idea that material reality is a subset of relationship patterns. Relationship is primary, objects are secondary, a gaze which has the potential to change the way we see the world and what we consider important. There is a similar worldview within traditional Celtic culture, known as the Web of Weird, a way of relating and seeing the world as primarily a web of agential relationships, with which it is possible to communicate in a way that is epistemologically impossible within a materialist cultural frame How can it be possible to talk to a rock or a tree if it has no brain or voice box? However, from a relational perspective it is possible to understand that even physical contact with a stone and our foot is an agential dialogue. So, there is a sense that the whole structure of the way people engage with nature and the world can be orientated completely differently by a different epistemology.

As the famous feminist biologist and philosopher Donna Haraway says, "It matters what stories tell stories" (Haraway, 2016, p. 35). In other words, the stories that we hear, the news and what we read and understand are based on our storying of other stories. To get a truly new perspective we need to try to get an overview, a meta perspective of what's going on. Haraway describes how we need to embrace what she calls "*tentacular thinking*," systems thinking which engages with the world like the agential, sensitive and mobile tentacles of an octopus. In other words, the kind of thinking that allows us to move beyond the merely logical and linear. A way of thinking that is constructed more like an organic animal form, like a spider's web, that not only connects with its environment, but is simultaneously sensitive to what it encounters. This is a very different way from how we usually think in Western culture.

We can explore this idea within our understanding of psychotherapy, using the ideas of trauma therapist, Deb Dana, who describes how, within trauma therapy, "*story follows state*." (Dana, 2018). In other words, what people say and how people articulate their view of the world, and themselves, does not have as much concrete meaning as we would like to suppose. In the same way that, from an Indigenous Aboriginal perspective, the material world is understood as a product of the dream time matrix, Deb Dana describes how our personal stories or narratives are the product of our underlying emotional states. She describes how what we experience in our bodies through the polyvagal system as emotion is brought to our sense-making through the mind by creating a story. In other words, our thoughts are mostly affect, or feeling in disguise.

What I'm suggesting is that we need to go back and reclaim an Indigenous worldview, not only opening our minds to become informed by contemporary Indigenous epistemologies, but also reclaim an understanding of our own Indigenous cultures, which we have lost or, to be clear about it, were stolen from us. But to be sure, we were all once Indigenous peoples and with this loss has come a marginalisation of both our attachment to nature and an accompanying deep systemic thinking.

This theme has been developed by Vanessa Machado de Oliviera in her book *Hospicing Modernity* (Machado de Oliviera, 2021), a very challenging book, but highly worthwhile. She describes a pathway to how we might be able to change our minds, or to change our story, by letting go of our addictive attachment to modernist thinking. Modernist thinking is thinking that relies on logical materialism and rationality, and a privileging of languaged epistemology as the only viable sense-making process. Machado de Oliviera makes the distinction between *wording* the world, that is, making sense of the world through spoken and written language systems, and *worlding* the world, engaging in reality through embodied relational practice (Machado de Oliviera, 2021). This requires breaking out of the usual Western conceptual framework, the difficult and unusual move from indexing and describing and thinking about the world, to engaging with the world as subject; the move from seeing the world as a collection of available objects, a consumer perspective, to bringing a better world into being by co-creating beauty. The idea here is that the role of humans on the earth is not as consumers but as producers, with the ability to create new realities. We are here to create beauty in the world and to give it away to others.

I experienced this difference strongly when I worked as a woodland manager. I already knew a lot about woodland plants as a biologist, but mostly in the abstract, I had learnt how to *word the world.* However, working in the woods every day for over a decade, walking through the woodlands and thinking about which trees to grow, which trees to cut, seeing the gradual seasonal change, I learnt to understand the relationships within the wood, and this understanding informed my work. I learnt how to *world the world*, how to move into a creative emergent relationship. I was able to embrace a completely different way of understanding the world. We see this approach to the world when we engage in work that is primarily about relationships, such as gardening, childcare, and art and craft work. In all of these activities, we are actually creating something, we are engaging with the world and creating a better world rather than just talking about it or describing it, or selling something.

We can also see this approach clearly described by the Aboriginal Indigenous scholar.

Tyson Yunkaporta, where the clear links between Indigenous thinking and deep systems thinking are evident. He says, "in our world (*the Aboriginal Indigenous world*), nothing can be known or even exist unless it is in relation to other things. Critically, those things that are connected are less important, than the forces of connection between them" (Yunkaporta, 2020, p. 149). In other words, it's not the things themselves that are important, It's the connection between. He goes on to say "we exist to form these relationships which make up the energy that holds creation together, and when knowledge is patterned in these forces of connection, it is sustainable over deep time."

We can see a connection here with the perspectives of quantum physics, in contrast to outdated Newtonian ideas, that atoms don't merely exist at some position within space and time, but space and time themselves are constellated by the relationships between the atoms (Barad, 2007). What we are seeing now is a much clearer articulation of the voices of Indigenous people within academia who are describing previously supressed ways of relating to the world.

One good example is the Mohawk seed keeper, Rowen White, who describes how she found herself working on an organic farm where she was asked to sort out some heritage vegetable seeds, tomato and squash, brought originally by settlers from Europe. This led her to consider her own Mohawk heritage, and whether she might be able to find heritage Mohawk maize seeds from her own lineage. Tracking down and learning to grow these traditional Mohawk seeds, she was able to reconnect with her culture through growing food and then building community around the harvest and cooking. Through the simple process of growing corn, she was able to reestablish her connection with her Mohawk cultural roots, a process she called an "*apprenticeship with the corn* mother" (White, quoted in Hemphill, 2021). She also believes that anyone can learn to reconnect to a deep sense of belonging through the process of growing seeds. She describes Western culture as the "*diaspora of the disconnected*," and that we can all reconnect deeply with our roots simply by growing food.

This brings us back to psychotherapy and attachment, and how it might be possible to regain a sense of belonging through processes such as growing food. However, this is quite difficult to access structurally, particularly for the cohort of people we might meet in psychotherapy who may not have easy access to land and land-based projects. The very structure of Western culture has separated us from the kind of attachment activities that were available to traditional and non-industrialised cultures.

An example of someone who is working to overcome this structural difficulty with urban Indigenous woman in the San Francisco, California is Corrina Gould. Corrina has set up a land trust of urban Indigenous women called Sogorea Te Land Trust (www.sogoreate-landtrust.org), which not only enables Indigenous people to grow traditional food but also restores and re-establishes a sacred relationship with the land through looking after the graves of ancestors and the practice of ceremony. Creating ceremony on land, which has an ancestral connection, is a process that Corrina refers to as re-matriation of the land, a process of reestablishing the relationship with land as the Great Mother, returning to our connection with the land as a way of healing our ancient ancestral bonds of attachment.

However, these approaches sit outside of a contemporary Western formulation of our relationship with the land and mental health. We can be helped to understand this by using different developmental mapping models, such as "*the circle of courage*" from the book Reclaiming Youth at Risk (Brendtro et al., 1990). This is a Lakota model that has been in use for about 15,000 years, but is still applicable, as are models from other Indigenous lineages, such as Gayaa Dhuwi (Proud Spirit) from Australia (www.gayaadhuwi.org.au), which is a 40,000-year-old Indigenous view of multiple interactive systemic attachment containers. These include the mind, the body, the family and community, and culture. What is meant by culture is growing food, making art, singing songs, the kind of culture that is not available by simply watching television. And finally, country, by which they mean nature and nature connection, and then spirituality, the spirit world and the ancestors. All of these are considered important components of Indigenous mental health, and to lose any of these is essentially an attachment issue. You can see from a Western perspective that we've lost quite a lot of these.

It's very interesting that these Indigenous cultures, now, after years of oppression, are able to come forward with these ancient ways of making sense of humans and nature. We can also learn from Indigenous wisdom about how we might reconnect with nature through a process of deep listening. This is described by Aboriginal elder Miriam-Rose Ungunmerr Baumann as a process called Dadiri, a way of communicating with nature through a process of deep inner listening, a quiet, still awareness, a tuning in to understand nature through beauty (Ungunmerr Baumann, 2021). This way of contacting nature is an aesthetic pathway to knowledge, and according to the Aboriginal people this practice is 60,000 years old. But it is also understood to be a practice that is available to all peoples of the world, including in the West, where knowledge of this approach to nature has been lost (Duncan, 2021).

So, it appears that Indigenous people have held knowledge and shared it, passing it down, and keeping a record of that through story, a tradition which we've lost in the West.

This Indigenous knowledge is also described by Malidoma Patrice Somé. In his book he describes his experience of being forcibly taken from his tribal village in West Africa to a Jesuit boarding school at the age of 7. Here he learnt to speak French and forgot his Indigenous language, but later returned to his village aged 20 to undergo a traditional West African Dagara initiation. He then continued his classical education in France and later in the United States, where he taught about his first-hand knowledge of West African spirituality and initiation. His conclusion echoes what has been described in other Indigenous lineages, that true healing comes "when an individual remembers his or her identity, their purpose chosen in the world of ancestral wisdom and reconnects with the spirit world." Malidoma is saying that healing is intimately linked with a deep sense of attachment that we can experience through reconnection with the spirit world, an idea that reaches well beyond the current Western paradigm of attachment (Somé, 1995, p. 38).

The perspective of Indigenous wisdom can also give us insight into the ecological crisis. Many different tribal lineages agree that the current meta-crisis needs to be taken very seriously and that we are, in their view, facing the end of the world as we know it. But they also hold traditional knowledge that the earth has already ended four times before. Indigenous cultures have maps and stories of how an end of the world might be navigated, and how cultures have managed to survive before. These Indigenous stories we now have available come to us from a knowledge system that forces us to push the boundaries far beyond contemporary Western knowledge of what we might mean by attachment and belonging. So, we come to the third practice, that of deep reconnection.

Reconnecting

This third practice I am naming as *Reconnecting to nature as the mothership*, and by that I mean nature as a place where we all come from. I am naming this as the *mothership* or nature as the *Great Mother*, because that enables us to think about nature in a way which is not usually used in the Western dialogue, but it articulates clearly a more ancient or traditional relationship where it is understood that we have come from nature, that we are all completely dependent on nature, and we go back to nature – that we *are* nature on every level. This position might enable us to engage with nature as an *ontological classroom*, in which nature has the potential to show us how the world actually works, if we have the ability to read what it is showing us. And in order to do this we need to let go of the old habit of objectification that's inherent in the Darwinian conception of evolution and embrace nature as subject, that nature is "subjects" all the way down, everyone is a subject.

This change of perspective often requires engagement with specific psycho-technologies, psychological and psycho spiritual practices, sometimes called "*soul work*" to facilitate a psychological systemic phase change, a process traditionally referred to as initiation. This is work I have been involved with for over 20 years, which started with my experience of contemporary vision fast or vision quest, ritual solo fasting in nature. (www.schooloflostborders.org, www.animus.org). This systemic eco-centric practice has been described most clearly by Bill Plotkin, who is able to outline these practices and how this can lead to what he calls *soul encounter* and *soul initiation*, a journey that he has mapped in his eco-centric development mapping tool, which looks beyond a merely psychological process and into an ecological process (Plotkin, 2021). This process was known by the alchemists as the great work, with a focus beyond psychotherapy, which leads to a deeper understanding of the world. The key transition in this process, described as *soul initiation*, leads from late-stage adolescence, a stage that Plotkin calls "*the wanderer in the cocoon*," often a stage where adolescents get drawn into psychological and social difficulties, and sometimes a state of being totally lost. Plotkin describes the next stage of soul initiation as "*the apprentice at the Wellspring*," where one can access their own deep source, our own wellspring and an accompanying sense of belonging. This work is well-established in the United States, Europe and elsewhere in the world, but is currently outside the framework and narrative of psychotherapy, although it is deeply therapeutic (Duncan, 2018).

The best way to learn about Ecosystemic psychotherapy is by engaging with psycho-technologies that facilitate work with the imaginal. A process which enables deeper encounters with nature and the soul beyond the limits of language. This process involves a letting go of the position of the ego and has been described best as "*dying practice*" (Little, 2001). This process can be supported through working with self-generated ritual in nature, which can help in the psychological preparation for an experience of soul encounter. This type of work has recently become more well-known through the emerging field of psychedelic-assisted psychotherapy. This kind of work that has been ongoing with some Indigenous culture for many centuries, where the idea of healing our sense of belonging has a different focus than Western psychotherapy. Within traditional cultures, the use of psychedelic medicine is in service of our relationship and sense of belonging to nature, rather than solely in service of our individual mental health. In this sense, it is an ecological process, a process of regaining a sense of belonging and a deep ecological sense of attachment.

What I would like to leave you with today is an invitation to be more curious about the roots of our sense of attachment and belonging, an invitation to learn how to recognise and engage with the complex imaginal systems both in nature and mind, an invitation to look beyond the traditional boundaries of attachment and to begin to reconnect with nature as our deepest and most primary sense of belonging.

***

References

Ainsworth, M. (1967). *Infancy in Uganda: Infant Care and the Growth of Love*. Baltimore MD: Johns Hopkins University Press.

Angarova, G. (2020). New and ancient story. *Animus Valley Institute* [Podcast]. Episode 53. www.animus.org.

Barad, K. (2007). *Meeting the Universe Halfway: Quantum Physics and the Entanglement of Matter and Meaning*. North Carolina: Duke University Press.

Bateson, G. (1979). *Mind and Nature: A Necessary Unity*. London: Wildwood.

Bowlby, J (1979). *The Making and Breaking of Affectional Bond*. London: Routledge.

Bowlby, J. (1990). *Charles Darwin: A New Life*. London: W. W. Norton.

Brendtro, L., Brokenleg, M., & Van Brockern, S. (1990). *Reclaiming Youth at Risk: Our Hope for the Future*. Bloomington, IN: National Educational Service.

Crittenden, P.M. (2008). *Raising Parents: Attachment, Parenting and Child Safety*. London: Routledge.

Dana, D. (2018). *The Polyvagal Theory in Therapy: Engaging the Rhythm of Regulation*. New York: W.W. Norton & Co.

Duncan, R. (2018). *Nature in Mind: Systemic Thinking and Imagination in Ecopsychology and Mental Health*. London: Routledge.

Duncan, R. (2021). Deep donkey and Dadirri: Asking Creatura out to play. *Murmurations: Journal of Transformative Systemic Practice*. 4(1): 32–47. https://doi.org/10.28963/4.1.4

Duncan, R. (in print). *Nature And Belonging. Navigating From Systemic Separation To Reconnection With Nature As The Great Mother*. London: Karnac.

Haraway, D. (2016). *Staying with the Trouble: Making Kin in the Chthulucene*. North Carolina: Duke University Press.

Hemphill, P. (Host) (2021). Seeds, grief, and memory with Rowen White. *Finding Our Way* [Podcast]. Series 2, Episode 6. https://www.findingourwaypodcast.com/individual-episodes/s2e6

Kripal, J. J. (2020). *The Flip: Who you Really Are and Why it Matters*. London: Penguin.

Little, Meredith (2001). The School of Lost Borders (personal communication). https://schooloflostborders.org/about/

Machado de Oliveira, V. (2021). *Hospicing Modernity: Facing Humanity's Wrongs and the Implications for Social Activism*. Berkeley, CA: North Atlantic Books.

Maldonado-Torres, N. (2004). The Topology of being and the geopolitics of knowledge. *City Analysis of Urban Trends, Culture, Theory, Policy, Action*. 8(1): 29–56.

Narvaez, D. & Bradshaw, G.A. (2023). *The Evolved Nest. Nature's Way of Raising Children and Creating Connected Community*. Berkeley, CA: North Atlantic Books.

Plotkin, B. (2021). *The Journey of Soul Initiation: A Field Guide for Visionaries, Evolutionaries, and Revolutionaries*. Novato: New World Library.

Schmachtenberger, D. (2020). War on sensemaking V. *Rebel Wisdom* [Podcast]. www.youtube.com/watchv=7Lqao

Somé, M.P. (1995). *Of Water and the Spirit: Ritual, Magic and Initiation in the Life of an African Shaman*. New York: Penguin.

Tippett, K. (Host) (2020, 4 June, updated 15 April 2021). On being, with R. Menakem. *Notice the Rage; Notice the Silence*. [Podcast].

Tye, L. (1998). *The Father of Spin, Edward Bernays and the Birth of Public Relations*. New York: Crown Publishing Group.

Ungunmerr, B. & Miriam, R. (2021). Miriam-Rose Foundation. www.miriamrosefoundation.org.au

Yunkaporta, T. (2020). *Sand Talk. How Indigenous Thinking Can Save the World*. New York: Harper Collins.

Chapter 4

Attachment and loss

How climate and eco-fiction can help us process our thoughts and feelings in a time of climate and ecological crisis

Maggie Turp

Part 1: A framework for thinking[1]

Introduction

Wendell Berry gave his volume of essays a question as a title: "What are people for?" (Berry, 1990) In response to that question, he writes of "… the hope of preserving qualities in one's own heart and spirit that would be destroyed by acquiescence." Related questions, such as "What is the nature of being human?" and "What enables people to hold onto their humanity in times of crisis?" are addressed in both psychotherapy-informed discussions and works of fiction. My aim in this chapter is to bring the two together in the service of helping us to think and feel our way in our relationship with the escalating climate and ecological crisis in which we find ourselves (see also Turp, 2023).

To begin with attachments, we humans become attached not only to other people but to "other-than-human" creatures and scenarios, including, but not limited to, landscapes, gardens, trees, the animals we observe as they go about their lives in the wild and the pets with whom we share our homes. Where there is attachment, there is inevitably a potential for loss. It is not for nothing that John Bowlby gave the title "Loss" to the third volume of his groundbreaking "Attachment and Loss" trilogy (Bowlby, 1969 [1991]). The climate and environmental losses impinging upon us at this point in our human history are numerous and serious. They include an unprecedented number of species extinctions, an ever-increasing number of homes and lives lost to fire and flood and record levels of displacement of both human and other-than-human beings. As children of the Holocene, we were not previously subject to events such as the 2025 flooding in Valencia that caused more than 220 deaths, or to UK summer temperatures of over 40°C, or to forest fires of the frequency and intensity we are seeing now. That extreme weather events will increase in number and intensity is no longer in doubt. At the same time, quite when, where and how any one country or region will be affected, and what the political fallout from the changes might be, is difficult to forecast with any degree of accuracy. The onslaught of losses combined with future uncertainties is

DOI: 10.4324/9781003733942-4

fuelling a pervasive sense of background anxiety, especially among young people (see Hickman et al., 2021).

A worsening situation combined with unpredictability is a sure recipe for profound unease. There are too many questions that cannot be answered except in terms of "perhaps" or "probably." Will the loss of agricultural land and the decline of pollinators lead to widespread food shortages? Perhaps. Will increased competition for food and water result in civil unrest, the breakdown of law and order and/or the eruption of new wars? Probably. When and where will such things come to pass? Impossible to say. Such is the current level of threat and uncertainty that we can no longer be confident that future generations will inherit a world where they can thrive.

Defences associated with existential threat

In response to the overwhelming evidence that the planet is heating, and heating fast, there has been a qualitative shift in the nature of the psychological defences most often in evidence. Where once the primary defence on display was denial – *nothing different is happening; there have always been climate fluctuations* – the more prevalent defence now is disavowal – *climate change is real, but we still have time; technological solutions will be found.* Where denial was openly argumentative and antagonistic, disavowal takes the form of disregard. We look away, retreating into an "as if" mode of functioning, continuing to operate as if we did not know what we know, in a largely unconscious endeavour to go about our business untroubled by the regular onslaught of disturbing news coming our way. In the language of attachment, we are in the domain of an avoidant attachment style, characterised by "… a profound belief that you have to stand on your own two feet, that seeking help is a form of weakness, that strong emotions should be avoided, particularly upsetting negative ones" (Music, 2019, p. 1).

Although an understandable response to the barrage of ever-worsening climate and ecological news, it would be a mistake to see disavowal as harmless. In 1912, the Titanic carried on across the Atlantic at full speed ahead, disregarding six separate warnings of unusually large amounts of sea ice. The consequences are well known.

The energy and aggression tied up in the defence of disavowal may be revealed when the defence is challenged. As an example, during one of several 2022 heatwaves in the UK, the meteorological office advised people to stay indoors during the hottest part of the day. Conservative MP John Hayes lashed out at those following the advice as "heatwave snowflakes," "cowards," and "everything that is wrong with modern Britain" (Stone, 2022). I recently attended a talk by Paul Hoggett, where he examined the phenomenon of disavowal and related it to values characteristic of the politics and economics of neoliberalism. In his talk, he referenced the way in which thinking can become split off from feeling, the normalisation of a crazy reality and the way in which cynicism and nihilism *annihilate meaning*, such that everything becomes a means to the end of profit, with no further justification required.

Psychoanalysis has long concerned itself with psychological defences, their function and the price we pay when they take over and run the show. In the late 1960s, a theoretical framework began to be developed by child analyst Esther Bick, which became known as "psychic skin theory." It provides an explanatory framework for the activation of particular kinds of defences that arise in the face of existential threat. On the basis of multiple observations of newborn infants and their mothers, Bick proposed that we are all subject at the start of life to a state of existential terror, except insofar as an attuned carer is available to keep that terror at bay. We are immobile. We can make no sense of the events unfolding around us. We are subject to whatever comes our way. We are helpless, and our very survival feels to be at risk (Bick, 1968).

Bick described how "psychic skin defences," the earliest and most primitive of the defences available to us, become activated in the face of existential threat such as that faced by an unprotected newborn. She described infants in the absence of an attachment figure tensing their muscles, holding their breath, clinging tightly to a blanket or focusing intently on a small detail of the environment in an apparent attempt to hold themselves together, to avoid the terrible fate of fragmenting, liquefying and falling into space: "When the baby is born he is in the position of an astronaut who has been shot out into outer space without a space suit … The predominant terror of the baby is of falling to pieces or liquefying" (Briggs, 2002, p. 66).

In a similar vein, Winnicott, a psychoanalyst and paediatrician working in the 1960s with babies and young children, described a state of "unthinkable anxiety," characterised by "acute confusion" and "the agony of disintegration" (1963 [1989], p. 90). These descriptions are attempts to put into words experiences that by their nature resist being encompassed by language. I think of all the World War I soldiers who survived trench warfare but never spoke a word on their return home of what it had been like – the bodies they had trampled over, the stench of the gases escaping from the dead and the agonised cries of their mortally wounded comrades.

Many of us who work as counsellors, psychotherapists or psychologists will have worked at times with people who have been violently attacked or sexually abused. In my experience, such patients often describe responses parallel to those observed by Bick. By focusing intently on one small detail of their environment, such as the pattern of the sofa they have been pinned down upon, they have been able to psychologically remove themselves from what is happening and, by extension, from their unbearable vulnerability and helplessness.

In *Sent Before My Time*, psychoanalyst and moral philosopher Margaret Cohen (2003) offers an eloquent account of psychic skin defences and disavowal as she reflects on her work as a psychotherapist in a neonatal intensive care unit. The NICU context is, of course, very different to that of the climate crisis. However, the two are linked by the experience of fear, guilt, loss and doubt with regard to survival. Cohen pays close attention to the defences that are activated by existential threat, reflecting on her own responses alongside those of others. She describes her efforts to observe a very premature baby, "Ewan," who develops meningitis and starts fitting. "I observed for a few more minutes, then could not bear it anymore. I looked around the unit, taking time off" (p. 13). At the same time, Ewan's mother, herself in a state of extreme distress, "felt that she could not look at Ewan anymore in case he was fitting" (p. 14). As everyone succumbs to the overwhelming urge to look away, Ewan is left to struggle alone in his pain and helplessness. Cohen discusses how the effort *not* to look away, to try to imagine and articulate Ewan's experience, militates against our natural desire to protect ourselves from mental pain.

> A common defence is to become angry, to blame someone and then to hate them – this eradicates the pain. Another is to withdraw into a cool analytic state, to be critical of the emotional mess everyone else is getting into and to take refuge in some cut-and-dried solution that denies the complexity of the situation. A third is to get on with your work – to put your head down and to refuse to engage. A fourth is to get into a rubbery, indifferent, or cruel state of mind where these things do not matter – they simply have no meaning.
>
> (Cohen, 2003, p. 158)

The reference to anger, blaming and hatred calls to mind the current escalation of anti-immigrant narratives, combined with physical attacks on their places of refuge. "Getting on with our work" is a form of disavowal available to all. Many of us, whether some of the time or all the time, respond to news of the current eruptions of violence with disavowal. We focus on something else – our work, or whether somebody has liked our latest post, or what we need to buy for dinner – so that we do not have to take on board the gravity of our situation. We "toughen up" so that we do not have to feel afraid.

Paul Hoggett references the phenomenon described by Hardin (1974) as the "politics of the armed lifeboat."

> Half a century later and Hardin's lifeboat scenario has literally come to life as wave upon wave of citizens from the Global South, uprooted by drought, flood, civil conflict and collapsed states, seek safety in Europe and North America. This has spurred the development of new forms of authoritarian nationalism, preoccupied with securing the nation's border against the waves of migrants.
>
> (Hoggett, 2023, p. 99)

When we turn away from the pain of our own vulnerability, we risk losing our capacity for compassion and thoughtfulness. It is only through remaining present to our predicament that we can hope to remain thoughtful and humane and give of our best as the crisis worsens.

The struggle to remain present in the face of a narrative of disturbing, sometimes horrifying, experiences is one familiar to many – perhaps all – counsellors and psychotherapists. It is in the nature of our work to offer ourselves as witnesses to client narratives that may include accounts of almost unthinkable suffering and loss, whether in the present or in a childhood scarred by abuse. We have as a profession developed an institutional support system in recognition of this challenge in the form of regular supervision. Supervision offers a confidential space in which to both think about what is going on in the client-therapist relationship and process the emotional impact of the profoundly disturbing revelations to which we are sometimes exposed. The climate listening circles that some climate-aware psychotherapists (myself included) are involved in can be seen as a kind of parallel, an attempt to mitigate the need for scapegoating or disavowal by providing a safe space to process the enormity of climate collapse and its actual and potential consequences. Climate and eco-fiction likewise offer a safe space for processing and reflection – of which more in Part 2. This is further enhanced when we come together to form cli-fi reading groups where we can share and elaborate on our reflections.

"Cold" and "warm" data

In 2018, academic Jem Bendell (2018) argued in a then-controversial journal paper that we are on a trajectory towards widespread civilisation collapse. Over the course of seven intervening years, a growing number of scientists have reached a similar conclusion. Several commentators have highlighted the fact that collapse is a gradual process rather than a sudden event. This process is evidently underway in countries particularly exposed to heat and drought, many of which are without the resources they need for mitigation. Some argue that the beginnings of collapse are also discernible in our own temperate and well-resourced country. Cost of living increases have left many people on low or middle incomes in the UK struggling to pay for essentials, with one in six UK households going hungry in 2024 and three in ten of the people referred to food banks coming from working households (Butler, 2025). When essential full-time workers, such as nurses, earn too little to feed their families, how long can things continue as they are? One immediately recognisable development is the rise of populist right-wing political views associated with the "armed lifeboat" states of mind referenced by Paul Hoggett.

A pressing question is how we might best sustain our humanity and capacity for thoughtfulness as conditions worsen. This are not the kind of questions that the cold data supplied by scientific research is best placed to answer. We can appreciate the soundness of the research. We can recognise that a forecast of a minimum of 2°C degrees of warming by the end of the century is extremely bad news. And yet it feels impossible to engage imaginatively with the barrage of facts and figures coming our way. What does a 2°C global temperature increase feel like? What does 12 metres of sea level rise look like? Gaia Vince (2023) has written about the large swathes of the globe that will become uninhabitable over the next 50 years, leading to an estimated 3.5 billion people having to seek new homes. How can we imagine a scenario involving this number of people on the move? Works of climate and eco-fiction – my focus in this chapter is on novels, but films, plays and television series have an equally important part to play – are a major source of warm data. Creating a narrative is how we make sense of our life experience; therefore, it is not surprising that stories speak to us in ways that statistics do not. Documentaries have their own part to play, giving us access to the present experience of individuals and communities already grappling with the impact of global heating. When it comes to future scenarios, however, only fiction can invite us into an experience – albeit a vicarious experience – of what potential developments might look like, what challenges they might present, and how we might feel and respond. And whereas a real-life crisis is likely to be traumatising and call for a rapid response, its fictitious counterpart offers a kind of "secure base" (Bowlby, 1988 [1991]), a safe rehearsal space in which to explore our thoughts and feelings and compare our imagined responses with those of the characters in the story.

Psychoanalyst Wilfred Bion (1962) emphasised the difference between "learning about" and "learning from experience." Research findings offer us an opportunity to "learn about" the current climate and ecological crisis. Stories offer us something different, the possibility of "learning from experience," as we live through a multitude of different scenarios in imaginative identification with invented characters, vicariously experiencing their struggles, their fears, their successes and their losses. It offers an opportunity to take on board and empathise with points of view other than our own and to mourn our losses in identification with fictional characters mourning theirs.

***

Part 2: Examples from climate and eco-fiction

Introduction

If the first part of this chapter has been in the spirit of "learning about," the remainder is intended as an invitation to "learn from experience," in this case the experience of reading and reflecting on a few selected examples of "cli-fi," a shorthand term for climate and eco-fiction. Although the term did not come into circulation until 2015, when it was coined by a journalist, Dan Bloom, the genre itself has a much longer history. We might think of John Steinbeck's 1939 novel *The Grapes of Wrath*, which brought the situation of the dust bowl and the resulting poverty, misery and desperation to the attention of millions of readers in the US and around the world. Or of Neville Shute's 1957 novel *On the Beach*, which follows the fate of a few remaining survivors of nuclear war in South Australia as they await the radioactive cloud that is heading their way. Each of these books had a profound influence on public perceptions and attitudes.

There is no one "correct" reading of a work of fiction and – as everyone who has ever been in a reading group will already know – one reader's response may be very different to another's. It is sometimes suggested that it is the work of literature to keep our hopes alive, to "offer a restorative spirit for the times we live in" or be a "tool to encourage change." A contrasting view, to which I wholeheartedly subscribe, is that the only work of literature is to engage our interest and thoughtfulness and elicit an emotional response to the characters and situations depicted. The rest lies in the space between the author and the reader. The "learning from experience" part of the chapter that follows consists of examples from some of the many works of climate and eco-fiction that have been important for me. After each example, I will discuss aspects of the work that I have found particularly thought-provoking and that have engaged me emotionally.

"We are where we are" by Kevin McCabe

One way of drawing a reader in is through humour, in particular satire. It is put to good use in the first example, a short story by Kevin McCabe, a member of Bristol Climate Writers. The story appeared in the journal *Dark Mountain* in spring 2019 and is republished here by kind permission of the author.

> When Leo opened his eyes there was something different about the sitting room. It was filled with smoke. It shouldn't have been a complete surprise as he had smelt smoke earlier in the day, but a cursory inspection had not disclosed its source, so he had decided to watch the football. Tedious defensive play from both sides had sent the game to a dreary nil-nil and him to sleep. In the meantime, the situation in his house had apparently deteriorated and ignoring it was clearly no longer an option.
>
> He rolled off the sofa onto the floor knocking over a half-filled can of lager, which was annoying, now he would have to clean the carpet as well. Better stay low, that seemed to be the thing to do, he had seen it in a film or something, and so he crawled over to the door. Fumes were streaming up from the basement utility room but the hall was otherwise clear so he stood up and went to the kitchen to get a bucket of water. He had a fire extinguisher under the stairs, but it was new and rather expensive. Better to wait for a real emergency.
>
> After a brief search, he remembered his bucket was catching a drip at the back of the toilet in the bathroom, so he went up the one and a half flights of the quirky three-storey building to get it. The bath taps thundered a good twenty litres of water into the bucket and he took it to tackle the orange flame that was now licking along the skirting boards in the hall. The liquid load arrested its progress and he returned upstairs for a refill.
>
> When he tried to descend again, the bottom step was glowing a healthy red, so he tipped the water over the bannister. It hissed and fizzled but did not make much of an impression. The lower storey looked lost.

We are where we are, he thought. As long as the fire stayed where it was, he would be OK, so he went into the bedroom to think about what to do next.

The afternoon sun poured in as he opened the curtains. Across the road he could see another house also on fire. In the window of the second floor his neighbour was sitting in his study watching television, which he found reassuring. In his own house, he realised the kitchen was now out of commission, but he didn't really need a kitchen to live, he thought, he could order takeaways. His plan to adapt to life one and a half floors up began to take shape.

No fridge was a bit of an issue, but when he was a student he had made one using a bucket of water and a large cloth. It was good enough for six-packs of beer, so it would probably do for milk and cheese.

The master bedroom would serve as his new living room and from now on he could sleep in the spare room, although he would have to move some of those boxes. In fact, he could throw them down the stairs and the fire would burn them.

OK, this was better, these were solutions.

He moved the double bed over to allow room to set up an office. He would have to work from home now that he couldn't get out. See, it wasn't all bad.

All the running around had made him feel sleepy. He assumed that was what it was, although the smoke made him a bit drowsy too. So he decided to take a nap. His acceptance that he had to move on from living at street level had taken all the stress out of the situation and he slept very well.

When he awoke it was dark. He found a can of lager under the bed, it was his lucky day, and he cracked it open. Looking across the road, he could see his neighbour still slumped in front of his television. He raised the can in his direction, sound bloke, he wasn't panicking, took a swig and went to the bathroom.

He stumbled to the toilet across the landing, now dense and black with smoke. When he flushed he realised he had no soap, so he wrote it down. There were bound to be lots of things he would need for his new elevated lifestyle.

Outside the bathroom, tongues of flame lapped at the top of the stairs and the carpet was smouldering. The fire was advancing faster than expected, which was a big disappointment. He had been thinking that the ground-floor fire was really a problem he could leave for the next owner. Well, he had a third floor; that was the advantage of these townhouses, what a shrewd purchase. So he gathered a pillow and his toothbrush and made his way upstairs again.

If the fire spread any further, he always had the attic. And, if he was honest, it could do with a clear-out.

Re-reading the story, I am reminded of the words spoken by Greta Thunberg in Davos in 2019: "I want you to act as you would in a crisis. I want you to act as if our house is on fire. Because it is." And yet it seems that this is an exceptionally hard thing to do. Leo repeatedly disavows the extreme seriousness of his situation. We understand, as he does not, that because he cannot face the loss of his home, he will lose his life. He lulls himself into a false sense of security, helped by an occasional can of lager and the sight of a neighbour whose house is also on fire and who seems similarly untroubled. He could know the reality of his situation – the evidence is clear as day – but he looks away. To acknowledge the true seriousness of his situation would be to accept the need for rapid, urgent change.

In *The Examined Life* (2013), psychoanalyst Stephen Grosz explores our resistance to change, particularly to major change imposed upon us rather than freely chosen. Drawing on examples from his clinical work, he describes it as "a passion for ignorance" (pp. 55–85). He relates the story of a patient, Francesca, who repeatedly describes incidents that indicate beyond any reasonable doubt that her husband is having an affair, while at the same time explaining them away. Grosz reflects that acknowledging the reality would require Francesca to make far-reaching changes, both external and internal. In a later chapter, "How a fear of loss can cause us to lose everything," Grosz cites examples of more extreme situations – as when, in September 2011, people in the south tower of the New York World Trade Center saw and heard a plane fly into the north tower and yet continued with telephone conversations or went into meetings. Grosz concludes that:

> We are vehemently faithful to our own view of the world, our story. We want to know what new story we're stepping into before we exit the old one. We don't want an exit if we don't know exactly where it is going to take us, even – or perhaps especially – in an emergency.
>
> (p. 123)

It seems that we will go to great lengths – conscious and unconscious – to interpret events in ways that reassure us that things will be OK, that major changes and their associated losses are unnecessary. We tend towards familiar patterns and pathways, those that have always worked in the past. As the saying goes, "Better the devil you know …" Leo's successive denials of reality are made palatable to the reader by exaggeration and a sense of the ridiculous. This is a fable and we can laugh at Leo's obviously nonsensical refusal to become alarmed. As an allegory, the story can help us to think about our own inaction in relation to our overheating/burning/drowning planet. Like Leo, we have placed our bets on not having to make major changes. Like Leo, we are now faced with a situation where rapid and dramatic change is required, where the cost is sky high, and where it is already too late to avoid major harm.

The High House by Jessie Greengrass

The High House is a work of climate and eco-fiction set in the near future. The story is narrated in alternating chapters by three protagonists – Caro, Sally and later Pauly, who is only a baby at the start of events. It deserves to be read in its entirety, and it is not my intention to reveal more than the necessary details of the unfolding action. My focus instead is on those parts of the book that revolve around attachments and disruptions of attachment in times of crisis (Greengrass, 2022).

The story begins with Caro, her father and her stepmother, Francesca, a climate scientist, living together as a family. They have a holiday home – the High House – in an isolated location on the coast. For 14-year-old Caro: "This was when it was still at the beginning of things, when we were still uncertain, and it was possible to believe that nothing was wrong bar an unusual run of hot Julys and January storms." For Francesca, however, the gravity of the situation is crystal clear, and she finds herself unable to empathise with those who have not yet faced up to the reality. Her frustration and exasperation with the stance of disavowal adopted by others make her hard to relate to. Caro comments: "Francesca didn't have the habit the rest of us were learning of having our minds in two places at once, of seeing two futures." Reading this passage, I thought long and hard about what it might mean to have one's mind in two places at once, about why it might be necessary and how it might be achieved. Do we need to intermittently enter into a state of disavowal, to look away, even as we know at some level that this is what we are doing, in order to be able to function?

Francesca and Caro's father have a baby, Pauly, but the ever more threatening climate situation results in their deciding to return to full-time climate campaigning. Caro describes Pauly's searing distress on the night of his mother's departure: "That night, Pauly wouldn't sleep. He stood at his bedroom door, his face wet with tears and sweat, and howled as though he were in pain." I found this passage excoriatingly painful to read. Folded into it is an important question: How do we square our individual and family responsibilities with our responsibilities to humankind in general and to the planet on whose resources we depend?

A strong point of the High House is the author's ability to represent complex and contradictory states of mind. Caro, on the one hand, resents being left to care for Pauly and on the other delights in him and his adoption of her as his primary attachment figure. When Francesca returns for a break from campaigning, Pauly rejects her. Caro reflects:

> I heard Francesca's hissed intake of breath. I heard her pause and walk away, and I felt a sudden spasm of guilt. How warm Pauly was in my lap, how comfortable, how soft! And how it must have hurt Francesca then to be in the next room alone and to have the truth confirmed that it wasn't that Pauly didn't talk at all but only that he didn't talk to her.

The morality or immorality of Francesca's decision to return to full-time campaigning is a recurrent theme. Later in the book, looking back, Caro reflects again on Francesca's decision that "she should place the general needs of the population above the specific ones of her family" but is unable to arrive at any lasting conclusion: "This is the absurdity of it, that I couldn't forgive Francesca because she chose the world over Pauly, and now I can't forgive myself because I didn't."

Not long after her parents' departure, Caro receives a panicked phone call from her father telling her to take Pauly and move with him to the High House, where Sally and her grandfather ("Grandy") are already in residence as caretakers. Caro succeeds, not without difficulty, in making the journey. Once there, she discovers that Francesca has equipped the High House with everything they might need for the foreseeable future, including a supply of medicines and clothes in different sizes for Pauly as he grows up. The garden is fertile and fruitful, and Grandy is a skilled grower of vegetables. The house is a prepper's dream come true, isolated and equipped with everything likely to be needed for the foreseeable future.

One of the strengths of the book is the way in which it undermines the fantasy that self-sufficiency and isolation offer a solution to collapse. We witness understandable difficulties in relationships between the various characters, in particular Caro and Sally. Caro rages against her abandonment by her father. Sally resents Caro's intrusion into her and Grandy's lives and is envious of her close relationship with Pauly. We are reminded that the ordinary, everyday difficulties of group dynamics do not go away in the face of larger challenges. Grandy is often called upon to smooth things over and emerges as a kind and compassionate figure who consistently offers containment when emotions run high.

In addition, both Sally and Caro struggle with survivor's guilt, which remains unresolved. Over the years, they see increasing numbers of displaced people passing by in the distance in search of sustenance and shelter and are faced with a conundrum: they know they should take people in; they *want* to take people in; but there are too many people in need; they cannot feed and house them all. Furthermore, giving away their location would threaten their own survival. Caro asks herself:

> What option is there in the end for the few of us who have survived but to be unforgivable and unforgiven. All those who might have lived instead of us are gone or they are starving, while we stay here at the High House, pulling potatoes from the soft earth.

Bradley Garrett's book *Bunker* (2021) consists of interviews with people – mostly well-off men – who have purchased a bunker or space in a bunker. While the interviewees see retreat to a bunker, should it become necessary, as an unproblematic response to collapse, Caro's reflections present a different, reality-checked conclusion: "We are only here because Francesca couldn't bear the thought of Pauly drowning. But the High House isn't an arc. We aren't really saved. We are only the last ones waiting."

For all the difficulties, living simply and close to nature is shown to have its rewards. Pauly is the character through whom the lyrical beauty of the landscape around the High House is most eloquently revealed. The house sits on the edge of a marsh leading down to the sea, and Pauly spends much of his childhood outside, falling under the spell of the wide, watery landscape and in close communion with the fellow creatures that share the terrain. Looking back, he reflects: "I remember the birds, and how it seemed to me that we hovered on the edge of understanding, and that they were only waiting for me to be still enough before they could come close to me and speak."

Time goes by, Grandy dies, and as the novel draws to a close, Pauly reflects on the isolation which can be his only end: "I think it will feel very cold. I think it will feel very empty. The last of us will not be buried." We are returned to the question of the meaning of survival in isolation, with no succession in sight.

Weather *by Jenny Offill (2020)*

Weather is a novel set in the present or very near future in the US. It revolves around the central character "Lizzie's" response to the situation in which she, and we, find ourselves. She lives with an awareness of the threat of climate catastrophe. She worries about her young son, Eli. Early on, her character is established by way of her interactions with customers at the university where she works, with her husband Ben, her son Eli and her troubled brother Henry. We get to know her as she lends an ear to the library users' stories, lets a customer in difficulties off a fine and rescues a wasp with a glass. These details establish her as a warm human being whom many of us would feel drawn to, making it easy to empathise with her worries and dilemmas. The library emerges as a small but important example of a "culture of care," as described by Sally Weintrobe (2021).

The story develops when Lizzie is invited to work for Sylvia, a climate and environmental campaigner, who produces a podcast called "Hell and High Water" and is much in demand. The climate crisis becomes the book's dominant theme, explored by way of comments and questions emerging at conferences, silent reflections and conversations between Lizzie and Sylvia and Lizzie and Ben.

What does, and what does not, constitute moral behaviour is a strong theme threading through the book. A particularly powerful passage calls to mind Maggie Cohen's writing on the loss of integrity involved in looking away from damage and distress. It articulates the hope that we will, in times of crisis, find the courage to bear witness to suffering, that we will retain our humanity by resisting the urge to look away. It is addressed by Sylvia to a conference audience, as she reflects on what it might mean to be a good person.

A moral person, she says, is calculated differently in times of crisis than in ordinary circumstances. She continues:

> Suppose you go with some friends to a park to have a picnic. This is, of course, morally neutral. But if you witness a group of children drowning in a lake and you continue to eat and chat, you have become monstrous.

One of my associations with this passage was an image that returned to mind of rioters stealing from an injured boy's backpack when riots broke out in the UK in 2011. I was reminded of how easily civilised behaviour breaks down in the face of chaos and crisis. The book is thought-provoking too in relation to the uneven impact of the climate crisis and the ease with which we in the global north look away from the sufferings of those in the Global South. As someone who feels little hope that we humans will find our way out of the crisis we have created, at least as far as material solutions are concerned, I have sometimes wondered why I continue to join demonstrations and to work voluntarily with the Climate Psychology Alliance. The answer is, I think, related to Sylvia's proposition. I am morally obliged to bear witness and speak out in the spirit of *not becoming monstrous*, of not standing quietly by while others suffer and die.

There are many examples in both life and literature of humour – particularly gallows humour – helping people to survive distressing situations, and yet humour is often in short supply in climate fiction. *Weather* is a welcome exception. We see how humour bonds Lizzie and Sylvia and helps them retain their positive energy and humanity. Lizzie has a strong sense of the absurd and finds the things people say at conferences a rich source for ironic comments:

- "One thing I'll say about it: lots of people who are not Native Americans are talking about Native Americans."
- "One thing that's becoming clear on our travels: people are really sick of being lectured to about the glaciers. 'Listen, I've heard all about that,' says this red-faced man. 'But what's going to happen to the American weather?'"
- Sylvia turns to irony to help herself manage the stresses and strains associated with her role as a public figure, commenting on the climate crisis:
- "I have to call you back," Sylvia tells me. "I'm about to send off this article, but I have to come up with the obligatory note of hope."

This latter example calls to mind a time when I ran a support group for climate scientists at a leading London research institute. The group members spoke of the burden of knowing how bad things really are, then the additional burden of having to inject a note of hope into their public messaging. Behind the strictures imposed upon these scientists – and indeed upon many of us – is the view that hope – the kind of hope that relates to things working out in the external world – is always a positive. According to this narrative, anxiety and dread push people in on themselves and into retreat. Hope, on the other hand, is an altogether good thing, associated with courage and positive action. In order not to spread despair, we are enjoined to keep the knowledge of the likelihood of collapse to ourselves. The enforced double-speak threatens to leave us feeling isolated and in bad faith. This is the shadow side of hope, the tyranny of the message to remain positive at all times.

The election of Donald Trump halfway through *Weather* significantly darkens the mood. "It's going to be too much," Sylvia said. "People who do this kind of work will break down, people will get sick and die."

Sylvia does, in fact, go through a period of quasi-breakdown, turning off her phone and retreating to a quiet place to restore herself. It has, as she feared, become "too much." The busy round of conferences and campaigns that she and Lizzie have been involved in grinds to a halt. As she becomes more isolated, Lizzie's state of mind also deteriorates. There is a sense of her falling apart, of lurching from one state of mind to another, unable to find a coherent narrative or course of action. She begins to fantasise about escaping from the ominous threats on the horizon. She talks to Ben about buying a gun, buying land somewhere cooler and developing survival skills. Sylvia responds with a stringent reality check: "Do you really think you can protect them? In 2047?," Sylvia asks. I look at her. Because until this moment I did, I did somehow think this. She orders another drink. "Then become rich, very, very rich," she says in a tight voice.

Disillusioned and yet unable to fully take Sylvia's words to heart, Lizzie becomes unsure of how to be or what to do: "I distract myself by staying up late, googling prepper things." She takes a break from inner turmoil by watching mindless TV – including a so-called "reality" show, *Extreme Couponing*. It is her self-deprecating brand of humour that signals a return to actual reality: "Then one day I have to run to catch a bus. I am so out of breath when I get there that I know in a flash all my preparations for the apocalypse are doomed. I will die early and ignobly."

Towards the end of the novel, Lizzie's brother's mental health worsens, and his partner throws him out. He moves in with Lizzie and Ben, sleeping on their couch. Lizzie's tolerance of her brother's frailties, her refusal to judge, her loyalty and lightness of touch are among the elements of the book that express the beauty and poetry present in everyday life. Holding on to the thread of living, regardless of what the future holds, is shown to be inherently meaningful, a stance of hope of a particular kind.

***

Creation Lake *by Rachel Kushner*

The last book for discussion is a more challenging option, challenging insofar as the central character, Sadie, is not a person most of us would wish to identify with. And yet there is no shortage of "Sadies" in the world today. Readers are no doubt aware of the stories revealed in the press of undercover police officers who infiltrated activist groups, and while there – and without giving any hint of their real identity or reason for being in the group – got into sexual relationships with group members, some of which resulted in the birth of children. How are we to imagine and understand the state of mind that underlies actions such as these? Can it be traced back to early trauma or deprivation? Or is it an expression of the neoliberal values of our times taken to their natural conclusion? For those who, like me, have asked themselves what kind of person it takes to act in this way, reading *Creation Lake* is a source of valuable insights.

Early on, Sadie reveals herself as a woman with a fake name, a fake place of birth, and – for the purpose of furthering her work – a fake intimate relationship. The sexual relationship is with "Lucien," a man who seems genuinely enamoured of her. She has drawn him in solely on account of his historical friendship with Pascal, the founder of the commune, which she is being paid to infiltrate. This relationship causes her some difficulties, but, notably, no moral dilemma.

> But these thoughts, my own actual thoughts, of Lucien breathing all over me, might remain. I might be stuck with the memory of him rolling me toward him at the hotel in Marseille, overpowering me and pinning me down in a manner he was convinced was right and good for both of us. Revisiting that, unable to block it, I wondered if I was getting paid sufficiently for this work.

Sadie's self-narrative is replete with envious, bitter attacks on others – the commune members – who have what she does not – values, places and groups where they belong, relationships where they can be themselves, and work that doesn't have to be hidden from the world. In her mind, she is the winner, above it all and doing just fine. Her cynicism and ruthlessness are revealed in a story from the past, when she recalls having tried to have a young man sent to jail for 20 years by entrapping him into buying fertiliser – ostensibly for a bomb. We come to understand that the commune members are in real danger: she will entrap them, regardless of what they may or may not have done, in order to be paid for her work. Paul Hoggett's talk, cited earlier in the chapter, comes to mind, in particular his reference to the way in which thinking becomes split off from feeling, a crazy reality becomes normalised, and the cynicism and nihilism involved *annihilate meaning*, such that everything becomes a means to the end of profit.

Sadie's comments about the commune are derisive, and for the most part we see it only through her jaundiced eyes – middle-class Parisians living on land that no sensible person would choose to try to farm, and grubby, ill-cared-for children kicking sand in each other's faces. By reducing them in her mind to worthless low life, Sadie justifies her persecution of them, a story all too familiar in real life. However, Pascal's wry and articulate response to being questioned by the police reveals a more serious intent and a thought-through philosophy that Sadie either cannot or will not see.

> "Sir, we hoe a row," he told the police. "We plant potatoes. We don't use pesticides. We nurture pollinators. But here is how the state does things: they have a deer population that's getting out of control, so what do they do? They bring in lynx. When farmers get upset about the lynx, the government introduces wolves. The wolves kill livestock, so the state makes it legal to shoot them. Hunting accidents increase, so they build a new clinic, whose

> medical staff creates a housing shortage, necessitating new developments. The expanding population attracts rodents, and so they introduce snakes. And so far, no one knows what to do about the snakes."

Another significant thread in the book takes the form of emails to Pascal – intercepted by Sadie – from a character, "Bruno," who has retreated to a cave and communes in his mind with its former inhabitants, the Neanderthals. This is a cli-fi – or perhaps an eco-fi – novel, and Bruno's is an original response to the escalating disaster – both general and worldwide and particular and local in the form of the drought affecting the region and the "solution" of water basins, which will provide temporary relief at the expense of disastrous depletion of the groundwater supply. Seeing no way forward, Bruno looks to the past and, in common with the people interviewed in *Bunker* (Garrett, 2021), is drawn to an underground location as a place of safety in a dangerous world.

Concluding thoughts

I began this chapter with some thoughts on a particular kind of hope, the hope of retaining our compassion and humanity in the face of existential threat. There is ample evidence to suggest that it will be no easy matter. The re-election of Trump in the US has been echoed in a sharp shift to the right in a number of European countries. In the UK, the far right is gaining ground and attacks on asylum seekers' hostels are escalating. Paul Hoggett, co-founder of the Climate Psychology Alliance (CPA), refers in his work to "the politics of the armed lifeboat."

> The ship of Western civilisation has hit an iceberg (the climate and ecological crisis) and is danger of sinking. As it begins to list heavily, passengers and lifeboats are flung into the icy water. Those who are drowning struggle to climb aboard the already full lifeboats but are beaten off by those on board, who fear their boat will sink if any more scramble on.

In the face of this onslaught on liberal values, we need more than ever to hold onto our capacity to feel, to think, and think about our feelings. I hope I have demonstrated some of the ways in which cli-fi and eco-fiction can help us to do so.

It is impossible in a single chapter to do justice to the many-facetted and increasingly impressive world of cli-fi. I have chosen to discuss a few offerings in depth rather than a larger number more superficially, a decision that has resulted in the omission of many works of climate and eco-fiction that would likewise reward thoughtful and extended reflection. Among those I have been more than a little sorry to omit is the *MaddAddam* trilogy by Margaret Atwood (2023), set in an imagined more distant future; *The Wall* by John Lanchester (2019) which conjures up Paul Hoggett's image of the armed lifeboat; *The Ministry for the Future* by Kim Stanley Robinson (2021), which imagines the creation of an international regulatory body for climate and the challenges it would face; *Blackout*, a thriller by Marc Elsberg, (2017) which traces the causes and consequences of an extended, Europe-wide power outage; *The Night Watchman* by Louise Erdrich (2020), a work of eco-fiction set in a Native American Indian community that invites us into a different way of seeing and being with the rain, the stars, our dreams and our fellow creatures; and *Venomous Lumpsucker* by Ned Beauman (2022), which explores the consequences of putting a monetary value on everything, including saving a species from extinction. The offerings out there are many, rich and varied. I apologise for those I have not found space to mention and hope that readers new to cli-fi feel encouraged to embark on their own personal journey of discovery.

***

Note

1 Note to reader: Please feel free to read Part 1 (Framework) and Part 2 (Examples) in whichever order you prefer.

References

Atwood, M. (2023). *MaddAdam Trilogy (Oryx and Crake, The Year of the Flood, MaddAddam)*. London: Virago.
Beauman, N. (2022). *Venomous Lumpsucker*. London: Sceptre.
Bendell, J. (2018, 10 August). Dialogue on deep adaptation. Available at: https://jembendell.com/2018/08/10/dialogue-on-deep-adaptation/
Berry, W. (1990). *What are People For? Essays*. New York: North Point Press.
Bick, E. (1968). The experience of skin in early object relations. *International Journal of Psychoanalysis. 5*: 558–566.
Bion, W. R. (1962). *Learning from Experience*. London: Karnac.
Bowlby, J. (1969 [1991]). *Attachment and Loss, Vol 3: Loss: Sadness and Depression: London: Penguin.*
Bowlby, J. (1988 [1991]). *A Secure Base: Clinical Implications of Attachment Theory*. London: Routledge.
Briggs, A. (ed) (2002). *Surviving Space: Papers on Infant Observation*. London: Karnac.
Butler, P. (2025). Failure to tackle dependence on food banks in UK driving public discontent. *The Guardian*, 10 September 2025.
Cohen, M. (2003). *Sent Before My Time*. London: Karnac.
Elsberg, M. (2017). *Blackout*. Richmond, VI: Black Swan.
Erdrich, L. (2020). *The Night Watchman*. New York: HarperCollins.
Garrett, B. (2021). *Bunker: What it Takes to Survive the Apocalypse*. London: Penguin.
Greengrass, J. (2022). *The High House*. Swift Press.
Grosz, S. (2013). *The Examined Life: How We Lose and Find Ourselves*. London: Chatto and Windus.
Hardin, G. (1974). Lifeboat ethics: The case against helping the poor. *Psychology Today.* 8: 38–43.

Hickman, C., Marks, E., Pihkala, P., Clayton, S., Lewandowski, R. E., Mayall, E. E., Wray, B., Mellor, C., & Van Susteren, L. (2021). Climate anxiety in children and young people and their beliefs about government responses to climate change: A global survey. *The Lancet, Planetary Health.* 5(12): e863–e873.

Hoggett, P. (2023). *Paradise Lost? The Climate Crisis and the Human Condition.* Winterhude: Simplicity Institute.

Lanchester, J. (2019). *The Wall*. London: Faber and Faber.

Music, G. (2019). Avoiding avoidance: neglecting emotional neglect and deactivated relationship styles. In Cundy, L. (Ed.), *Attachment and the Defence Against Intimacy*. (pp. 1–22). London: Routledge.

Offill, J. (2020). *Weather*. London: Granta Books.

Shute, N. (1957). *On the Beach*. London: Heinemann.

Stanley Robinson, K. (2021). *The Ministry for the Future*. London: Orbit.

Steinbeck, J. (1939). *The Grapes of Wrath*. New York: The Viking Press.

Stone, J. (2022). Tory MP says "cowards" taking precautions in heatwave are "snowflakes" and everything wrong with Britain. *The Independent*, 15 July.

Thunberg, G. (2019). *No One is Too Small to Make a Difference*. London: Penguin.

Turp, M. D. (2023). Imaginative engagement with the climate crisis: The role of climate and ecology fiction. In Aspey, L., Jackson, C., & Parker D. (Eds.), *Holding the Hope: Reviving Psychological and Spiritual Agency in the Face of Climate Change* (pp. 111–121). Monmouth: PCCS Books.

Vince, G. (2023). *Nomad Century: How Climate Migration Will Reshape our World.* Flatiron Books.

Weintrobe, S. (2021). *The Psychological Roots of the Climate Crisis: Neoliberal Exceptionalism and the Culture of Uncare*. London: Bloomsbury.

Winnicott, D. W. (1963 [1989]). Fear of breakdown. In Winnicott, C., Shepherd, R., & Davis, M. (Eds.), *Psychoanalytic Explorations*. London: Karnac,.

Chapter 5

Secure bonds

The intersection of art, attachment and environmental challenges

Catherine Jacobs

Better together

Figure 5.1 Interconnections 1. Limited edition photographic print.

Climate mitigation is an overwhelming, complex and urgent crisis that is not just a future threat, but is already happening across our global ecosystems and economies. We are living with extreme weather events, the loss of biodiversity, and increasing financial and social upheaval. We need to ask, "Will our actions be enough?" and not "Will my actions be enough?" state Solnit et al., (2023);

DOI: 10.4324/9781003733942-5

"This is a communal quest in which everyone can bring their talents, visions, desires, access – and if one person struggles, we can help each other up" (Lutunatabua, 2023, p. 15). As the effects of climate change are often gradual, they can be difficult to fully understand, especially on an emotional level. Psychological distance can be involved, as we see it as a problem that affects other people and the future, rather than something that impacts us as individuals in the present (Dunlap, 2013). We also have natural cognitive biases that lead us to prioritise immediate concerns over long-term threats, making it even harder to engage with climate mitigation (Korteling, Paradies, & Sassen-van Meer, 2023). Fossil fuel industries and their allies feed into this with strong financial incentives to delay climate action and misinformation campaigns to sow doubt, and politicians may downplay the crisis to avoid making unpopular economic decisions.

There is a call for psychologists to improve the integration of psychological concepts into feasible interventions for the climate (Nielsen et al., 2021; Nielson et al., 2024). With its focus on the importance of strong affective ties, Attachment Theory has much to offer for the understanding of how we can build the necessary community support and joint action to make meaningful progress. The theory explains why we look for physical and emotional closeness in key people in our lives, particularly when we are feeling distressed, fearful or threatened. John Bowlby argued that attachment is not learnt, but is an innate system – the infant needs to be cared for, and the parent or caregiver must respond to the infant's needs for the attachment to be successful and adaptive. Bonding behaviours, such as proximity-seeking and crying, evolved because they increase an infant's chance of survival by keeping them close to a person or people who provide food and protection. Our attachment patterns learnt in these early caregiving experiences form the groundwork for how we show up in our relationships, and in the group and collective settings we are part of, influencing how we interpret each other's behaviours. If primary caregivers can provide comfort, security, and protection, they are termed "safe havens." As we grow up, we develop "internal working models," which are internalised mental representations that we hold about ourselves and others (Bowlby, 1973). They also shape how we communicate our emotions, how we handle conflict, and our ability to trust others and be vulnerable (Fonagy & Target, 2002). Secure attachment, whether from childhood or "earned" in adulthood through supportive relationships, therapeutic or organic, is strongly linked to more effective emotion regulation strategies, while those with insecure attachments may experience more difficulties in managing their emotions and connecting with others. These models and patterns are not fixed and can change across the lifespan (Bifulco et al., 2004).

Traditionally, attachment research has focused on the parent–infant bond, but the theory has also long been applied in the context of adulthood. In everyday life, those with more security have higher levels of trust and take opportunities for the confiding of feelings when faced with emotionally stressful situations, which are alleviated if our attachment figures can offer attuned care and comfort. Attachment insecurity, at its heart, is feeling that our world is not safe or trustworthy and means we spend more time in survival mode, in avoidance or ambivalence, or both if we are more disorganised in our strategies. Attachment frameworks can be seen as part of the picture, helping to predict how we respond to uncertainty; coping with climate mitigation is best done in relationship to others. When our psychological frameworks support relational, human interconnection, we are more likely to develop a willingness to act resilient, which is important knowledge for those applying Attachment Theory in practice and clinical settings. The ASI has a careful, strengths-based feedback protocol which, in utilising the person's verbatim accounts from the interview, is usually well-received. There can also be strong psychoeducational benefits for the person interviewed, which can include a better understanding of any barriers to closeness (moving away from any hypothetical thinking of how good support networks could be in the future), an understanding of their resiliencies and risks and a wider knowledge of attachment in general.

More recently, I qualified as a Senior Practitioner Executive and Leadership coach, choosing to train in Bowlby's legacy at London's Tavistock Centre. Coaching conversations using the systems-psychodynamic Tavistock model concentrate on the loss and anxiety that may well be showing up in management roles and wider organisational systems undergoing change, and work with "beneath-the-surface," unconscious phenomena, such as countertransference and projections (Armstrong, 2005). This approach, in which we offer longer sessions to allow responses to emerge, is informed by psychodynamic theory, group relations theory and systems thinking and is underpinned by Attachment Theory. I adhere to Peter Hawkins's definition of coaching as a "collaborative inquiry, where both parties are fully engaged, but neither has the answer and the agenda is set by the wider life and context of the coachee" (cited in Whybrow et al., 2023, p. 149). Attachment Theory also informs my coaching practice in how I'm careful to model secure attachment principles and develop the psychological safety and trust needed to work in-depth with my clients dealing with turbulent "VUCA" environments. on behalf of each other and the climate. This involves accepting human co-accountability, having a readiness to act together, and caring for others, both present and future generations (Nisa et al., 2021). Empathy is thought to be innate, but is also modelled, as more securely attached children show greater kindness and generosity (Music, 2014). In adulthood security is linked to increased prosocial behaviours such as giving more to charity (Cundy, 2017). How we engage with climate mitigation, of course, reflects our broader attachment strategies, our capacities for trust, connection, reciprocity, and care. And just as with human relationships, when our attachment to our planet feels fragile or unsafe, we might respond with anxiety and avoidance, or detachment and denial.

I attended the 27th Annual John Bowlby Memorial Conference *Attachment, Climate Crisis and the Natural World* (2024) with great interest. To meet climate challenges, we need to find better ways to come together, to harness our differences and relate differently with each other, "paying close attention to the quality of connection and conversation between individuals, groups and entire organisations in order to develop their effectiveness, agility and responsiveness today's businesses, societal and global changes call for" (Cavicchia & Gilbert, 2018, p. 5).

The business world now commonly uses the acronym "VUCA" (Bennis & Nanus, 1985) referring to "volatile, uncertain, complex and ambiguous" organisational environments. A "VUCA" world is one where change happens rapidly, patterns are harder to predict, systems are too complex to analyse, and there are too many "unknown unknowns." Leaders, whether in business or other sectors, are often faced with working with only partial information, not knowing what to do next, but nevertheless needing to act, which can bring with it associated stress, increased reactive emotional states and burnout. Bringing ideas and methodologies from research psychology, leadership coaching and a visual art practice, in this chapter I offer thoughts on framing the psychological dynamics of uncertainty to help to foster trust, bonding and connection between us in the context of working collectively with climate change.

***

My interdisciplinary self

Over my research career I have specialised in in-depth, semi-structured interview assessments of attachment style, parenting, life events and early experiences of care and abuse, working for many years with colleagues at the Centre for Abuse, Trauma and Suicide Studies, (CATSs), now at Middlesex University, London. My attachment expertise is in the application of the Attachment Style Interview for adults and adolescents (ASI), designed by Professor Antonia Bifulco (Bifulco et al., 2002). The measure forms part of a systemic methodological model that continues the work of professors George Brown and Tirril Harris, who advanced psychological understandings of mental illness by operationalising life stressors and vulnerability factors in community populations, highlighting the interaction between environmental stress and individual susceptibility in the development of psychiatric disorders (Brown & Harris, 1978).

The ASI is an example-based, behavioural interview that "echoes Bowlby's principal concern with observable behaviour rather than speculations about the inner world" (Oskis, 2019, p. 25). The measure aids understanding of a person's current supportive network, including investigator ratings of the quality of their partner and other close supportive relationships. It identifies their ability to make and maintain relationships; attitudinal barriers people can have to relating, and produces an overall attachment style. The ASI is the only measure that differentiates between secure and insecure attachment styles, not only in terms of *how* insecure a person is (Mildly, Moderately or Markedly), but what *type* of insecurity they show (Anxious-Enmeshed, Anxious-Fearful, Avoidant-Withdrawn, Avoidant-Angry). Secure styles, or those at only "Mild" levels of insecurity, and those rated Moderately Withdrawn show high levels of resilience.

Since childhood, I have been sustained by a strong interest in connecting to the natural world, and this remains the basis of my visual art practice that runs alongside my academic and consultancy career. Both in and outside the studio, I make landscape-based works that aim to shift our perspectives to better engage with our wider world. Open-ended and ambiguous in their nature, my artworks also have a psychological side. Functioning a bit like "Rorschach" prompts, I present them as invitations to explore uncertainty, inviting our playful engagement with what we are seeing in front of us, in the here-and-now. I situate art and aesthetics as mediums for curiosity and in connection to the whole range of our human experience. As with a coaching context, if something speaks to us and we have the desire to engage with it, we are much more likely to be able to achieve change. Engagement with the arts gives us a chance to take us out of our own heads and visualise the world as others see it. Lorenzo Servi says that art reminds us, "That there is not one singular way of doing and seeing things … we can have more than one solution, and questions more than one answer" (Servi, 2016, p. 107).

Wearing my interdisciplinary hats, this is a subjective account, where I present concepts and techniques from my fields that I hope will be of use for facilitators of arts groups incorporating climate work in their remits. My aim is to offer ideas and practices for working with uncertainty and ambiguity that will help increase our prosocial behaviours, grow authentic connections with others and challenge cultures of disconnection. All the artworks shown are mine.

***

The function of love

Figure 5.2 The function of love, 1. Limited edition photographic print.

When we participate in tasks together, or engage collectively within groups, teams or wider organisations, our attachment patterns interplay between us. Employing theories and practices from attachment frameworks aids our understanding of the interpersonal, social and political contexts of how groups and organisations operate. This mostly happens unconsciously as we all bring with us the internal working models we have for ourselves and our relationships, which are shaped by our early experiences. These frame what we expect from others (such as "People always leave" versus "People are dependable"), how we deal with relationship problems (such as "I am not lovable" versus "We're just having a rough patch") and our sense of self-worth in love (such as "I deserve respect" versus "I need to prove I am worth loving"). And, of course, this has implications for how we might work collectively to combat the climate emergency. Barbara Fredrickson reminds us, "In a world replete with threats, uncertainties, and ceaseless distractions, the urge can be strong to simply look out for oneself" (Fredrickson, 2013, p. 161). Attachment-based clinical psychiatrist Daniel J. Siegel has long emphasised that children need the "four Ss" of being "Safe, Seen, Soothed and Secure" to aid healthy development. This applies equally to how we function as adults, and these attachment principles can be incorporated into guidelines for those developing or leading arts or activism groups. More recently, Siegel has widened his thinking to focus on how our Western, consumerist cultures promote our identities as isolated and separate from each other. He argues for the uptake of much wider relational thinking, and for us to better recognise that "the bodily, inner self is connected to other people and to the planet, the whole of nature" (Siegel, 2022, p. 8). Adopting a broader relational and systems perspective enables us to comprehend the interdependent nature of the whole world, thereby empowering both individual and collective action toward the creation of a just and environmentally resilient world for all.

Alongside his clinical work, John Bowlby was also politically active throughout his lifetime and career, engaging with the changes in social policy that were much needed after World War II and the rise of a more collectively oriented social-democratic policy. Mayhew (2006) writes of the importance of Bowlby's political life, which has been much less reported than his development of Attachment Theory. "Bowlby could conceive of love as a natural potentiality within children that could serve wider social ends" (Mayhew, 2006, p. 20). Bowlby was a prominent member of the New Fabian Research Bureau,[1] working towards democratic socialism, and in his work with evacuation schemes, he highlighted the need for homely settings for the children as crucial to keep their anxieties about their enforced parental separations at bay.

We all need to be seen and understood. Attuned attachment bonds lead to ongoing patterns of security and autonomous behaviours that benefit not just us, our families and us as individuals, but society in general. Researching the impact of love and connection over a long career, psychologist Barbara Fredrickson (2013) emphasises that working with positive emotions can broaden our mindsets and build our resourcefulness. We have greater activity in the brain regions indicating increased perceptual breadth when we are experiencing authentic connections (Schmitz, De Rosa & Anderson, 2009; Algoe & Frederickson, 2011). Attachment-based psychotherapist Sue Gerhardt brings attention to the isolation and lack of care that the neoliberal politics of today bring with them in her book "Selfish societies: How we all forgot to love one another and made money instead" (Gerhardt, 2010). Psychoanalytic psychotherapist Sally Weintrobe also addresses "cultures of uncare" in two seminal books that focus on the context of climate change engagement (Weintrobe, 2013, 2021). In them, she argues that to be treated with greater care, we must first see people as having subjectivity and being in need of care. This can then lead to a "collective reset in values, systems, and ways of living rooted in ecological integrity, social justice and compassion" (Centre for Compassionate Leadership, 2025).

In *Art as Therapy* (2013), Alain de Botton and John Armstrong identify seven psychological functions of art: Hope, Sorrow, Remembering, Understanding the self, Growth, Appreciation and Rebalancing. They argue that art plays a therapeutic role by helping us process our emotions, better connect with ourselves and others, and find meaning in our lives. In the context of climate action, this healing dimension becomes even more vital. Participating in artistic activities can allow us to transform our private anxieties into collective understanding, fostering empathy, solidarity, and resilience in the face of environmental crisis. Engagement with art can help us recognise that suffering is a normal part of the human experience and offers space for expressing our grief over ecological loss. At the same time, art can inspire hope by presenting visions of beauty, joy, and a more sustainable future, which is especially important when we might feel overwhelmed by negativity. It can also encourage personal and collective growth by expanding our emotional and intellectual horizons, exposing us to new perspectives and experiences that might otherwise remain beyond our reach.

***

Processing loss and grief

Figure 5.3 Rural Idyll 1. Photographic print & jigsaw.

Writer and activist James Baldwin stated, "Not everything that is faced can be changed, but nothing can be changed until it is faced" (Baldwin, 1962, BR11). Grief is an adaptive response to loss, which includes the loss of social bonds, of autonomy, of identity and of certainty, as well as the loss of a loved one. Bowlby (1980) believed that suppressing grief can lead to ill health, and when we lose someone we are bonded to we activate our attachment system to cope with the trauma. The Bowlby Centre has long highlighted the importance of facilitating mourning to develop security "We see (mourning) as vital to the working through issues of abandonment, emotional misattunements or neglect as well as trauma and abuse". Undertaking a grieving process therefore lets us recognise and accept our emotions, easing the path towards healing and recovery. Over time, we can learn to thrive in the present, rather than stay overwhelmed.

Attachment and psychoanalytic models of grief and loss are equally applicable to working with climate change. A person's response to loss is, of course, shaped by their cultural practices and social expectations as well as their inner working models and beliefs. Working through losses and disappointments is crucial to the formation of a secure sense of self and the capacity to form and sustain intimate relationships. Those with secure attachments may process grief more adaptively, while those with insecure attachments might struggle more with their emotions. Wilfred Bion's (1962) theory of containment is useful to consider, as this refers to our ability to convert intolerable states of mind to more tolerable ones within a relational interaction whereby one person (the container) receives, processes and returns the intense and overwhelming feelings and experiences of another (the contained). Professor Kristen Neff, in her research on the theory and practices of self-compassion, emphasises that the essential part of this process is working with people who have a genuine intention to help. Self-compassion is composed of mindfulness (being self-aware in the present moment), self-kindness (being understanding and supportive towards oneself when experiencing pain, failure, or inadequacy) and crucially also involves leaning into our common humanity. This means that, rather than feeling isolated when we are struggling, we can choose a more embodied understanding by acknowledging that we all suffer at times and this is part of our shared human experience (as well as that of many other sentient beings). When practised, components of self-compassion have been shown to significantly enhance interpersonal connections, greater relationship satisfaction and attachment security by fostering empathy, reducing isolation, and promoting healthier relationship dynamics (Neff & Davidson, 2016). We cannot take away other people's grief, but we can find ways to support them as they process it.

We can also apply learning from the business world to our understanding of how teams and groups deal with loss and change. Harry Levinson, who, in the 1950s, applied psychoanalytic theory to the field of organisational management, wrote, "All change is loss, and all loss requires mourning" (Levinson, 1972, p. 80). Organisations experiencing change often fail to support their staff in processing feelings of loss, leaving individuals to handle these emotions on their own (Fragouli & Tourlaki, 2020). Leaders of today require personal qualities such as high levels of emotional intelligence, self-awareness, vulnerability, empathy and compassion to bring their teams with them in the ever-changing workplace (Brown, 2015). Approaches such as the Tavistock's systems-psychodynamic methodology working specifically and relationally with the "beneath-the-surface" loss that may be present in our roles and in our complex group systems can help. By studying how businesses manage loss and change, and through the practical application of psychological theories, climate groups and activists can gain valuable insights into building stronger, more resilient, and more impactful movements for social change.

Our feelings of climate grief and ecological anxiety will likely mirror our individual attachment wounds, where healing will be best undertaken relationally with people we trust and feel safe with. While the climate emergency can create life events such as catastrophic flash floods, overall we experience its effects as much more of a continuous and ongoing process, which therefore presents unique challenges for community grieving. Losses are not just in the past, but also in the here-and-now and are often projected into the future. We can, for example, feel anticipatory grief (the emotional process of grieving a potential loss) and transitional grief (a sense of loss and longing for what was, or what could have been). We can come together to discuss "What exactly it is we are mourning"? with psychotherapeutic-based climate thinkers such as Rust (2020), Randell (2005, 2009), and Weintrobe (2013, 2021), all positioning eco-grief as an adaptive and healthy response to the climate emergency, best worked through in relational contexts. "The frameworks of care now needed must include social support to work through the damage to our environments, our societies and our minds" (Weintrobe, 2021, p. 298). For them, developing rituals to process grief and anxiety is absolutely core to us coming together to mitigate the climate emergency and successfully decarbonise. Leaning into our suffering with self-compassion and a sense of care can help us unite against the powerful political propaganda supporting business as usual.

The Climate Psychology Alliance is an international charity that counteracts cultures of uncare by facilitating people to meet in community listening circles with careful, safe psychological containers in place. Their online journal[2] is a useful read, with issue 7 (May 2025) offering insights and thoughts about the importance of connection and cultivating community. Engagement with art and the arts is another way to direct our actions, and at the same time deal with the complex feelings evoked as we face global environmental challenges. Art does not have to be a solitary activity; group-based art projects, public memorials, installations and performances work with the power of people coming together. If facilitation frameworks embrace psychological safety, principles of containment and attachment security, then we can also use some of these sessions to reflect on loss, become less overwhelmed and give us agency to act. "The arts, by their very nature, are strong sensory inputs that engage multiple systems … they are uniquely capable of accessing interconnected neural pathways that allow us to process emotions" (Magsamen &

Ross, 2023, p. 32). One example of a global storytelling project is Amir Nizar Zuabi's "The Herds," a large-scale theatrical production of animal puppets that in 2025 "migrated" from West Africa to Scandinavia. The project aimed to spark climate change awareness through emotional sensory experiences in everyday environments. "The minute you add beauty into the equation, it becomes personal, it becomes yours. It becomes something worth fighting for" (Zuabi, 2023).

***

Working with difference

In systems-psychodynamic executive coaching, we provide a framework for leaders to understand and manage the complex and often unconscious dynamics related to difference in the workplace. This approach delves into the underlying emotional and psychological processes that impact organisational culture, interpersonal relationships, and leadership effectiveness. Our individual attachment styles form part of this picture. As we all bring our attachment patterns to the teams we participate in, these become part of what happens between us when we undertake tasks and interact together. The groups we gather in will usually therefore incorporate a range of attachment styles (secure, anxious-ambivalent, avoidant and disorganised). People who are more securely attached are comfortable with intimacy and independence. They tend to be trusting, have a positive view of themselves and others, and can balance closeness with autonomy in their relationships. People with more anxious-ambivalent strategies are often led by a desire for intimacy alongside a deep fear of abandonment. They may experience high levels of anxiety when interacting with others and can be overly dependent on people for reassurance. Avoidant people are usually less comfortable with emotional closeness and tend to very highly value their independence. They suppress their emotions and may distance themselves from others, often appearing self-sufficient. People who are disorganised in their ways of relating will display a mix of anxious and avoidant behaviours and may well exhibit unpredictable shifts between clinginess and distance.

Our attachment styles will be activated if we find ourselves stressed and holding difficult feelings brought on by any severe life events or difficulties that may currently be happening in our lives. This will also likely apply if we choose to engage with the enormity of the climate emergency – our usual patterns of relating will be in the room with us. Working with the arts is one way we can find safe outlets for processing difficult experiences, reducing anxiety and grounding ourselves in the present moment. A substantial body of research now consistently shows that there is a strong link between engaging in creative or artistic activities and improved mental health outcomes (Jean-Berluche, 2024). A well-facilitated group can provide a safe haven where, over time, members feel secure enough to be vulnerable and express their true selves.

While they can sometimes bring interpersonal challenges, our differing ways of relating greatly enrich the teams and groups we are part of, but they need to be met with sensitive, empathetic leadership responses. When managed well, our different ways of relating offer unique perspectives for everyone involved and can elevate creativity, resilience and collaboration in any group or team. Arts facilitators would not be expected to understand the specific best practices of working with a range of attachment styles, but some simple knowledge from Attachment Theory and research could be helpful. One idea would be to align people's roles within groups with their strengths, for example, more avoidant people may shine in solo projects and more anxious ones in collaborative roles. It is beyond the scope of this article to discuss in-depth how to lead people with different attachment styles, but the following paragraphs offer a few thoughts to enhance mindful and empathetic facilitation.

In group settings, securely attached people may be naturally more inclined to seek out ways to act, as they might be better able to bear, and discuss, the emotionality of asking questions such as "What is my contribution to this?," "How willing am I to be changed?," "What needs to change that I can campaign for?" Applying this to collective movements, securely attached individuals could be effective at encouraging collaboration and enhancing mutual support across groups, especially if there is a lot of change, with new members regularly dropping in to sessions. Securely attached people will usually have good abilities to manage their own emotional responses, alongside those of others. If someone has avoidant or anxious attachment insecurity, they may well be defended against emotions such as anger, sadness, or fear and may not have the vocabulary to directly express these feelings. Engaging with the arts, or art making, can give those emotions a symbolic form, allowing them to externalise what might otherwise be buried and giving insights into feelings that were not fully realised before. Working with the arts can allow us to reframe how we think about our climate narratives to situate them as more imaginative explorations of new ways to live and come together.

Avoidantly attached people often value feeling in control, so focusing on the process of creating art, rather than on the final product or outcome, can make the experience more manageable for their defences. Facilitators can consider how the creative process for people with these relational patterns may well take more of an analytical or cerebral slant and be aware that they may not like to engage with their worries or anxieties. Someone with an avoidant attachment style might also be more likely to disengage from the emotional aspects of climate change altogether, perhaps because it feels too overwhelming for them, or because they do not know how to process the enormity of the problem. In contrast, they might be able to engage with climate change mitigation in calm, clear and practical ways.

More insecurely anxiously attached people might experience high anxiety and will likely appreciate help to find hope amid the chaos and uncertainty of climate work. Group facilitators can be mindful that anxiously attached people may not be able to pinpoint how they feel in a given moment, and they therefore may well have fears of being misunderstood or can behave with high sensitivity to others' reactions. When faced with the challenges of climate change action, an anxious individual could easily become overwhelmed by the scale of the problem and experience intense fear about the future. They may also feel powerless, thinking their personal actions cannot make a difference. Working to make collective art, or engaging in the wider arts together, may help counteract this.

Insecure-disorganised attachment is characterised by a mixture of behaviours and strategies that are often ineffective or conflictual in terms of relational proximity-seeking and real social support. A strong mistrust of others will often be an issue for them, and this will likely come to play in the groups they participate in. A person with disorganised attachment patterning might feel drawn to connect with others, but will also likely be deeply afraid of being hurt or abandoned and feel out of control at the same time. Facilitators will need a clear understanding that, for this person, who may present very emotionally at times, being in a group setting can evoke much anxiety about whether they are in a safe space, and there is the potential for confusing and chaotic dynamics in relationships as the person struggles to regulate their feelings.

***

Security frameworks for the arts

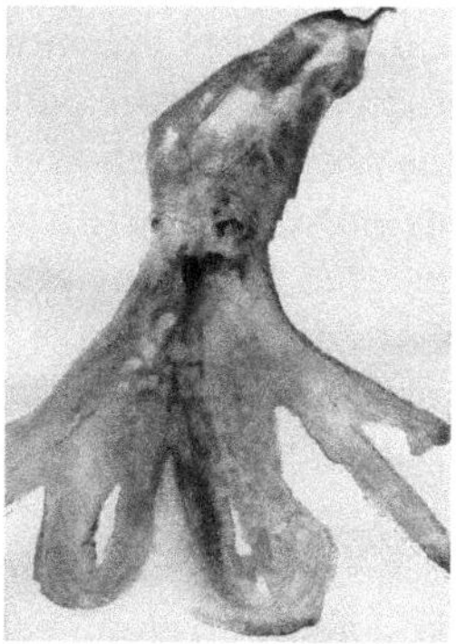

Figure 5.4 More-Than-Human 8. Watercolour on paper.

The arts offer a huge range of practices that involve creative expression and storytelling. Engaging with the arts can help us bond and heal, especially when we feel safe, seen and cared for. Engaging diverse communities is essential for effective climate action, with group organisers needing to try as much as possible to remove any physical, financial and psychological barriers to the participation of new audiences. Interactive activities, such as art-based projects, can help to turn passive viewers into active participants in the climate conversation, especially if the work involves sharing local stories that connect broad climate issues to people's immediate surroundings. There are several pathways into arts-based climate activism. We can join or set up climate groups, running creative or family-based activities. Art galleries and museums often run workshops or short courses as part of climate or nature-themed exhibitions, as do some national arts organisations that work directly with environmental issues. Nature-based arts groups, with their role to conserve and protect environments, will also provide ways for us to play a part in helping our local ecosystems. Urban and rural open-air festivals are also well-placed to invite people to respond to their local natural landscapes. There is guidance for working in this way, with useful resources available, including the Gallery Climate Coalition's artist tool kit,[3] Culture Declares Emergency[4] and the British Council's report "Mapping trends and best practice in climate action and sustainability in the arts."[5]

Understanding and applying concepts from Attachment Theory, psychodynamic thinking and best codes of practice from the coaching world, I believe, can help those facilitating arts practices form safe and creative spaces for connection and exploration. Holding others' emotions and employing secure attachment principles are skills that can be learnt and developed through practice. It is inevitable that our projections and defences will arise when we work with hard-to-face climate issues, and facilitators of groups working with the arts and the climate emergency require sensitivity to ensure people do not feel too threatened by their involvement. Linda Cundy (2019) reminds us that our core patterns of attachment are default settings which we return to when we feel stressed and are where our defences are most evident. Unless we are in denial, eco-anxiety is likely to activate our core patterns and defences. Harvard's Amy Edmondson first used the phrase "psychological safety" (1999) to explore its relationship to how teams grow and perform. This describes the shared belief, within teams or groups, that it is okay and safe to talk about ideas or concerns, make mistakes, ask questions, and take risks without worry of negative repercussions. When psychological safety is active in groups, a foundation of trust and mutual respect builds, which empowers each member to contribute their insights and collaborate across disciplines and perspectives effectively. When an environment fosters this interaction, groups will expand their abilities to innovate, solve problems, and adapt to change. When this is not present, our attachment strategies and patterns of reacting to threat are likely to be activated. Facilitators crucially also need to be emotionally aware and skilled at managing their own emotional responses to the group dynamics and individuals within it. This capacity is known as reflective functioning or mentalising and involves the ability to work with and imagine one's own and others' mental states, including thoughts, feelings, desires, and intentions, and to understand how these influence behaviour (Fonagy & Target, 2002)

***

I have explored how Attachment Theory is a fundamentally adaptive model, as our strategies are responses that we develop to help us navigate specific social and emotional environments, often originating in our childhood experiences. Understanding our attachment patterns helps us to make sense of why we can't always connect and be open to support or ask for help (Bifulco et al., 2009). When individuals do not feel threatened in relationships, an activation of their caregiving system is much more likely to be facilitated. Ways of offering safety include the modelling of secure base or safe haven behaviours, attending to Siegel's "Four Ss," and understanding the benefits of the application of wider aspects of Attachment Theory, such as individual differences in emotional regulation. This can, alongside professional supervision and governance, help group leaders offer consistent, clear and careful codes of care and ethical frameworks. The charity Action for Happiness has designed "Five Golden Rules" underpinning sessions, which are very applicable to arts-based groups. These are:

> We really listen to each other. Practicing giving each other our full attention and valuing the perspectives, experiences and insights that everyone brings | We are open about our feelings. Noticing them and practicing being open about them, recognising that being a little vulnerable in this way helps us all to connect | We focus on what is working well | We practice safe boundaries | We maintain a spirit of kindness.
>
> (Action for Happiness, 2025)

Arts-based groups create opportunities for emotional expression, but members may experience fear of being judged or may feel anxiety about whether their contributions are viewed as good enough. In environmental activism groups, common forms of anxiety would likely include eco-concerns, signs of burnout or frustration over perceived setbacks. Individuals new to the community will often worry about how much they belong and whether they fit in. Group contracts and boundaries provide psychological safety in managing dynamics and help establish safe and containing spaces with clear expectations, and increase a sense of ownership. These help to promote effective collaboration within the group and increase the likelihood that the group will achieve its goals. Group agreements also foster a culture of trust and open communication, promoting effective teamwork and collaboration. Clearly defined roles and responsibilities, along with agreed-upon protocols, can help minimise misunderstandings and conflicts within the group. Furthermore, contracting allows for flexibility and adaptability to changing circumstances, ensuring the group can adjust its approach as needed.

Secure attachment behaviours bind us in our shared humanity and bring about a sense of psychological connectedness. Recent findings from Claudia Nisa et al. (2021) argue that this connectedness appears to be of use in tackling climate change as the attachment system overcomes any ingroup favouritism and helps develop empathy. Their research working with priming attachment security stimuli indicates that by increasing empathy for humanity in general people can be influenced to care more about climate change. Priming works by activating mental representations in memory, making them easier to access and likely to influence our future thoughts and actions. Their findings also show it is possible to work with politically conservative people who might be more sceptical (Nisa et al., 2019, 2021). In his book on climate psychology, Mathew Adams (2016) offers guidance that would also be useful for group leaders to deepen their thinking. And for those wanting a very comprehensive dive into the methodologies around the emerging field of Climate Coaching, I would recommend the Climate Coaching Alliance's book *Ecological and Climate-Conscious Coaching* (Whybrow et al., 2023) and Cox and Flynn's book *Climate Change Coaching: The Power of Connection to Create Climate Action* (2022)

***

From proximity-seeking to awe-seeking

Figure 5.5 Liminal 1. Limited edition photographic print.

Pioneering activist Rachel Carson stated that "The more clearly we can focus our attention on the wonders and realities of the universe about us, the less taste we shall have for the destruction of our race" (Carson, 2022, p. 86). In this chapter, I have discussed how a crucial element of climate mitigation is about us coming together in innovative and creative ways, and how concepts and practices from Attachment Theory, such as the secure base, and the practical application of learning from psychotherapeutic and coaching methodologies have much to offer. Weintrobe (2013) explains how easily our psyches can split and become divided – caring deeply for what and who is "near us," while feeling less connected to far-off communities, lands, and futures. Engagement with the arts can serve as a powerful medium for bridging psychological and emotional distances, fostering empathy, and cultivating a broader sense of relationality with distant communities, environments, and imagining possible futures. Through reflective practices, the arts can expand our capacities to connect with experiences beyond our immediate context. My art practice is rooted in the aesthetics of landscape and shaped by a deep sense of nature connectedness, which is the personal, emotional experience of feeling intimately linked to the natural world. I encourage, in people viewing my artworks, a heightened sensitivity to the natural environment, fostering moments of reflection, presence, and reconnection, and exploring the psychological dimensions of our relationship with nature, inviting viewers to engage more fully with the wonder and awe present in our everyday lives.

Understanding how we are intrinsically linked to nature can also help us to recognise our interdependence with the Earth's ecosystems we depend on. The effects of nature on human emotions and mental well-being have long been studied, with evidence showing that spending time in nature can reduce tension, enhancing affective states and satisfaction in life (Nejade et al., 2022). The dominance of city living, while giving us the wonders of culture and innovation, tends to isolate us from the ecosystems that sustain life. Studies on nature connectedness are useful to examine here. Recent research stresses that to really connect with nature we should be moving away from generalised ideas of spending time "escaping" in nature and towards a greater emphasis on specific activities of engagement that we can undertake in places and communities local to us (Christens et al., 2025). The activities we choose are recommended to be simple to carry out and encourage more mindful practices, such as listening to birdsong, noticing wildlife, or smelling flowers (Richardson et al., 2021). Values that emphasise the interdependence of human communities and nature can be usefully termed relational environmental values. Part of this is the concept of psychological oneness, which is defined as the belief that people and the natural world are one and involves a reduced focus on the self as an individual (Diebels & Leary, 2019; Coomber & Harré, 2022).

Over the last decade or so, there has been a substantial increase in research on the experiences of awe, positioning them as psychological states or mechanisms that can increase our nature connectedness. A white paper[6] (Allen, 2018) presented evidence showing that if we actively seek out and experiment with experiencing this emotional state, our sense of separateness can start to diminish, and we can even decrease our materialistic behaviours. "Awe serves a vital social function: it expands an individual's perspectives from narrow self-interest to others' needs and collective concerns" (Piff, Singhal, & Bai, 2025). Although awe and wonder may feel like personal, internal experiences, there is mounting evidence to suggest that they have wider "we-minded," social effects as well, specifically on our levels of kindness and generosity.

> Awe is typified by two states – vastness and accommodation, or *wow* and *whoa*. Vastness is when something, be it an incredible vista or complex idea … strikes us in a way that makes us feel small and connected to the broader ecosystem of humanity. As our mind wraps itself around this new, moving experience and accommodates it, our schema changes, as does our view of the world. Because of this *small self*, awe makes us more prosocial, meaning we want to be better to each other through emotions like compassion, empathy and generosity.
>
> (Parker, 2023, p. 94)

A recent paper entitled "Awe is a conduit to cohesive collectives" will be of particular interest to groups working with climate change (Piff et al., 2025). The authors emphasise that utilising this emotion increases cooperation and our affiliation with working collectively. This is an important finding as to function well in groups, we need to think less about our individual needs and more about the needs and customs of the wider group. Research is even suggesting that engaging with awe experiences can increase our openness to differing viewpoints, which is essential to working with diverse groups of people (Stancato & Keltner, 2021). There is also evidence that the action we may or may not take for the climate is strongly influenced by our social norms, and awe may motivate us behaviourally if we believe others are acting in similar ways (Prade & Saroglou, 2023).

In her book *The Power of Wonder* (2023), Monica Parker looks in-depth at what wonder actually is, the science behind it, how to practice it, and provides many ways of using it across daily life. She positions wonder as having five dimensions: "watch" (openness), "wander" (deep curiosity), "whittle" (absorption), "wow" and "whoa" (awe). Awe can be defined as a strong feeling of respect or amazement brought on by something that is beautiful or sacred to us, and can at times be a positive or negative experience. As an artist working with the psychological, I have long been interested in this emotion, agreeing with American curator Denise Markonish's assertion that "Wonder is not passive; we can cultivate it … treat it like a practice" (Markonish, 2022, p. 51). The default mode network is the area of the brain that is associated with wonder, and interestingly this is also the area associated with theory of mind. This is the ability to understand that other people have their own thoughts and expectations. It is a crucial social-cognitive skill that allows us to build relationships, communicate and be understood, and cooperate with others. Secure attachment, characterised by a sense of safety and the availability of caregivers, is generally associated with better theory of mind development (Fonagy & Bateman, 2016). This experience is echoed in relationships; if we feel the comfort of safe havens in our inner circle, we can better explore the world and then return to be soothed.

Professor Dacher Keltner (2023), a leading authority on awe research, suggests we cultivate practices such as "awe walks." When we deliberately set aside time to slow down and wonder, we can try to imagine with fresh eyes the vastness of things or lose ourselves close-up in the detail of a plant. Artist, activist and author Sophie Howarth's anthology of essays, poems and photographs (2025) offers us creative practices to help us recover our senses of wonder and awe. With my love of the photographic, I was especially drawn to her "Lightstalking" exercise that invites us to "Watch the sunlight as it moves into, through and out of your home over the course of the day. Notice it … slipping through blinds or flooding around the edges of curtains" (Howarth, 2025, p. 40). "Glimmers" are micro moments in our day that we can nurture to spark a sense of joy and safety in us (they can be seen as the opposite of triggers, which evoke feelings of fear or danger). Research from the application of Polyvagal Theory (Porges, 2022), which proposes that the autonomic nervous system, particularly the vagus nerve, plays a crucial role in determining our physiological and behavioural responses to perceived safety and threat, is showing that experiencing glimmers can help cue our nervous systems to feel safe and regulated (Dana, 2018). Seeking them out or recognising when they happen can shift our nervous systems from states of agitation or feeling numb or disconnected, to ones of greater connection and calm, and engaging with the arts can help us achieve this. I try to seek out small moments of connection with everyday beauty to nurture and inspire me.

Innovative methodologies which engage our imaginations are also presented in the late Joanna Macy's environmental work (Macy & Johnstone, 2022). These include participating in "Councils of All Beings" that join us in community with each other and our local more-than-human multi-species worlds. I recently attended an "Interspecies Council," facilitated by UK-based Moral Imaginations, speaking as "ladybird" and debating with "lichen," "fox" and "plane tree," to name a few. The workshop enabled us to playfully discuss problems facing our local ecosystems from the point of view of the ecosystem and its wildlife, bringing non-human beings into political decision-making. Also in the UK, established environmental arts organisations Invisible Dust and Julie's Bicycle provide thought-provoking events and consultancy. In times where we have so much information available at the click of a button, engaging in simple, local nature activities and seeking out emotions of wonder and awe can bring us together, help us look beyond our immediate day-to-day concerns and adopt the longer-term perspectives needed for us to engage with the emergence of solutions to climate change and other global challenges.

***

Embracing uncertainty in art

Figure 5.6 Uncertainties 22. Limited edition photographic print.

In my art practice, I work with positive aspects of ambiguity, presenting deliberately open-ended, uncertain artworks to invite and encourage people to engage more with their own subjectivities. Embracing uncertainty playfully and creatively can offer us new perspectives, insights and ways of being as individuals and collectives.

Donald Winnicott (1971) is well known for his writing about the nature of playing and its relationship to psychological well-being and development. For him, play promotes the possibility of a new experience for the person, with emphasis on what is experienced in the here-and-now or, if it's relational, in the transference in therapeutic settings. The key is in bringing to our minds and bodies something that was hitherto unknown. Wilfred Bion's concept of "without memory and desire" refers to the psychoanalyst's ideal state of mind during a session – approaching each encounter with a fresh perspective, free from preconceptions based on a desire for specific outcomes or past experiences (Bion, 1967). Complementary to the Tavistock model (Armstrong, 2005), organisational researchers highlight that knowledge is increasingly positioned as arising from interaction and emerges from the complex interplay between individuals and their environment (Simpson & French, 2006). This perspective emphasises the dynamic and interconnected nature of knowledge creation, where shared experiences and social interactions play crucial roles. Leaders in "VUCA" environments need to have the capacity to step back and form thoughts about the emotional matrix of their complex businesses – understanding what the system (group, team, wider organisation, stakeholders) needs at that time, akin to a "good enough" parent "knowing" what their baby requires of them.

Engagement with art is another method that we can use to surprise ourselves. Writer and activist Rebecca Solnit reminds us in the catalogue for the exhibition "Dear Earth" that "Art can inform, but it has a larger task in the climate crisis, and that is to make us the people we need to be to respond to the crisis in the ways we must" (Solnit, quoted in Fowkes, Fowkes, & Solnit, 2023, p. 27). She also speaks that practical hope "Is not about what we expect. It is an embrace of the essential unknowability of the world, of the breaks with the presence, the surprises" (cited in Mason, 2014, p. 154). The state of not knowing or leaning into the unknown is crucial for artists as the creative process often involves a strong sense of direction with more contemplative states of exploration, experimentation and play. I am an artist whose approach is organic, often developing from "accidents" that occur as part of the process. In my practice I regularly use insights from the book *On Not Knowing: How Artists Think* (Fisher & Fortnum, 2013), which could also be a useful resource for arts and climate facilitators. Other artists work better by visualising a completed piece of work, which they then set out to create.

I work mainly with the medium of photography, as I find its inherent ambiguity – it always needs further context to show what is really going on inside and outside the frame – can be used to increase our engagement with emergence and the as-yet unknown. I use the techniques a camera offers, such as working with a macro lens and blur, to present uncertain artworks as vessels, not as something to be analysed and fully understood, but for enhancing playfulness in the here-and-now of the encounter between the viewer and the work. Siegel (2022) also offers photographic metaphors in his work around the relational, integrated self.

> Close-up, we sense the body is the centre of experience; wide-angle, we sense our relational connections with people, and wider still, with the whole of nature, the whole of the universe; and taken together, we can fill the full range of our inner and relational centres as the interconnected self.
>
> (Siegal, 2022, p. 241)

"Uncertainties" (2002–2003) is an early photographic series of mine that was staged in the studio by making small-scale temporary installations out of everyday beauty materials, including face creams and body lotions, just for the camera. These, when enlarged, become uncanny, psychological landscapes, lacking a sense of place or narrative. In solo exhibitions[7] of the series, I included an audience participatory element, inviting people to share their experiences encountering the work with each other. Responses to "Uncertainties 22" (Figure 5.6) covered both the playful and profound:

> I am not what they say I am not | Vanilla Ice Cream | Surfing the storm | The final thought at the end of time | Night of the sea monster | Swan running in a pond | Hope | Darkness forever | Confusion | Oil slicks | Absorbing absorption | Hot silver | Man running away from life's responsibilities |

> Distorted face trapped | A single sperm! | A cat being drowned | An alien being born | The first encounter | Bird on a branch in winter | Been here before | Hopeful, light at the end of the tunnel | Whirlpool | Someone crying on a book | Rain ripples | The devil in a lake of sulphur blowing smoke rings | Hot silver | Something is being made alive |

Encouraging participants to playfully engage with the unknown and discover new possibilities could be a useful strategy to embrace when working with climate issues. Rosen (2019) has published many engaging play-based ideas and practices. I especially like his brief "Emotional transfers" group exercise (Rosen, 2019, pp. 123–124), where we can act out an emotion, such as anger, alongside another person acting out a different one, such as sadness, and then move these to each other. A few other ideas to incorporate include experimental prompts to encourage diverse interpretations and fostering an environment where group members can share their ideas and approaches, and build collaboratively on each other's discoveries. Facilitators can provide opportunities for their group members to try new techniques or materials without fear of failure, and at the same time can encourage people to pay attention to any unexpected moments and insights that may arise. Working in this way is all about finding innovative, unthreatening ways to help people better cultivate their state of mind to increase their care and actions taken for the environment.

***

Concluding thoughts

Across our lifespans, our attachment patterns, internal working models and behaviours often lie beyond our awareness, but nevertheless influence how we communicate our emotions, handle conflict and engage with difference. In our everyday lived experience, attachment is all about relationships and interactions, social support and help seeking – and explains our ability to trust, be vulnerable, or depend on others. This chapter has argued that global change happens through intimate dialogue and connection between us. It is in the repatterning of our behaviours, in the changing of where we place our attention, and the healing ability of each of us, that will allow us to better engage with other people, including future generations, and our wider ecosystems. Working with climate change and decarbonisation requires a willingness to live with complicated feelings, and to do this we need to engage our imaginations and embody a sense of shared responsibility and care. Stemming from Attachment Theory, these characteristics are associated with secure ways of relating, which can be positioned as core principles to help groups achieve meaningful change. Embracing practices of not knowing from systems-psychodynamic coaching, and art can be playful, engaging and helpful.

In recent years, there has been a clear "relational turn" or shift across many academic fields to emphasise the importance of relationships and highlight the interconnectedness of human beings, our social structures and systems. This thinking moves us away from viewing people as isolated entities and focuses on how connecting with others is fundamental in shaping our identities and sense of self. Ecology involves examining the links between organisms and their surroundings, and the word is rooted in the Greek for house/dwelling (*oikos*) and study of (*logia*). This chapter has offered ideas from different disciplines to show how the incorporation of key attachment components, such as secure base, safe haven and security in the context of arts facilitation and groups, could help us get our house in order, work better collectively to deal with the challenges of climate change, and strengthen our investment in our local and global environments.

***

Notes

1 Now the Fabian Society.
2 *Explorations in Climate Psychology*.
3 artist-toolkit.galleryclimatecoalition.org
4 culturedeclares.org/take-action/culture-takes-action-toolkit
5 www.britishcouncil.org/sites/default/files/mapping_climate_action_and_sustainability_in_the_arts.pdf
6 *Science of Awe* (Allen, 2018).
7 Midlands Art Centre, Birmingham (2005) and the British Psychological Society's Annual Conference, Dublin (2008).

References

Adams, M. (2016). *Ecological Crisis, Sustainability and the Psychosocial Subject: Beyond Behaviour Change*. London: Palgrave MacMillan.

Algoe, S.B., & Frederickson, B.L. (2011). Emotional fitness and the movement of affective science from lab to field. *Am Psychol.* 66(1): 35–36.

Action for Happiness (2025). Available at: https://actionforhappiness.org (accessed 10 August 2025).

Allen, S. (2018). *The Science of Awe*. A white paper prepared for the John Templeton Foundation by the Greater Good Science Centre at UC Berkeley.

Armstrong, D. (2005). *Organization in the Mind: Psychoanalysis, Group Relations, and Organizational Consultancy. Occasional Papers 1989–2003*. London: Karnac.

Attachment, Climate Crisis and the Natural World (2024). *John Bowlby Memorial Conference, the Bowlby Centre*. London, UK, 20 April.

Baldwin, J. (1962). As much truth as one can bear. *New York Times* Book Review, 14 January, p. BR11.

Bennis, W., & Nanus, B. (1985). *Leaders. The Strategies for Taking Charge*. New York: Harper Row.

Bifulco, A., Moran. P.M., Ball, C., & Bernazzani, O. (2002). Adult attachment style I: Its relationship to clinical depression. *Social Psychiatry and Psychiatric Epidemiology.* 37(2): 50–59.

Bifulco A., Figueirido B., Guedeney N., Gorman L., Hayes S., Muzik, M. et al. (2004). Maternal attachment style and depression associated with childbirth: Preliminary results from a European/US cross-cultural study. *British Journal of Psychiatry (Special supplement)*. 184(46): 31–37.

Bifulco, A., Moran, P., Jacobs, C., & Bunn, A. (2009). Problem partners and parenting: Exploring linkages with maternal insecure attachment style and adolescent offspring internalizing disorder. *Attachment & Human Development.* 11(1): 69–85.

Bion, W.R. (1962). *Learning from Experience*. London: Heinemann.

Bion, W.R. (1967). Notes on memory and desire. *Psychoanalytic Forum.* 2: 271–280.

Bowlby, J. (1973). *Attachment and Loss. Vol. 2: Separation, Anxiety and Anger*. New York: Basic Books.

Bowlby, J. (1980). *Attachment and Loss, Vol. 3: Loss, Sadness and Depression*. NY: Basic Books.

Bowlby Centre Website, thebowlbycentre.org.uk (accessed 15 June 2005).

Brown, G.W., & Harris, T.O. (1978). *Social Origins of Depression: A Study of Psychiatric Disorder in Women*. London: Tavistock.

Brown, B. (2015). *Daring Greatly: How the Courage to Be Vulnerable Transforms the Way We Live, Love, Parent, and Lead.* New York: Avery.

Carson, Rachel (2022). Remarks made on acceptance of the John Burroughs medal for excellence in nature writing, 2 April 1952, New York. In Lear, L. (Ed.), *Lost Woods: The Discovered Writing of Rachel Carson*. Edinburgh: Canongate Books.

Cavicchia, S., & Gilbert, M. (2018). *The Theory and Practice of Relational Coaching: Complexity, Paradox and Integration*. London: Routledge.

Center for Compassionate Leadership (2025). Newsletter, August. https://mailchi.mp/00d9d64c9210/the-everyday-path (accessed 16 August 2025).

Christens, B.D., McCormick, Megan L., & Wolf, Joshua (2025). Understanding and promoting nature connectedness: A human ecological perspective. *Current Opinion in Psychology. 65*: 102046.

Coomber T., & Harré N. (2022). Psychological oneness: A typology. *Rev Gen Psychol.* 26: 49–67.

Cox, C., & Flynn, S. (2022). *Climate Change Coaching: The Power of Connection to Create Climate Action*. London: Open University Press.

Cundy, L. (2017). Fear of abandonment and angry protest: Understanding and working with anxiously attached clients. In Cundy, L. (Ed.) *Anxiously Attached: Understanding and Working with Preoccupied Attachment* (pp. 19–48). London: Karnac.

Cundy, L. (Ed.) (2019). *Attachment and the Defence against Intimacy: Understanding and Working with Avoidant Attachment, Self-Hatred, and Shame*. London: Karnac.

Dana, Deb. (2018). *The Polyvagal Theory in Therapy: Engaging the Rhythm of Regulation*. New York: W.W. Norton & Company.

De Botton, A., & Armstrong, J. (2013). *Art as Therapy*. London: Phaidon Press Limited.

Diebels, K.J., & Leary, M.R. (2019). The psychological implications of believing that everything is one. *J Posit. Psychol.* 14: 463– 473.

Dunlap, R.E. (2013). Climate change scepticism and denial: an introduction. *Am. Behav. Sci.* 57, 691–698.

Edmondson, A. (1999). Psychological safe learning behavior in work teams. *Administrative Science Quarterly.* 44(2): 350–383.

Fisher, E.A., & Fortnum, R. (Ed.) (2013). *On Not Knowing: How Artists Think*. London: Black Dog Publishing.

Fonagy, P., & Target, M. (2002). Early intervention and the development of self-regulation. *Psychoanalytic Inquiry.* 22(3), 307–335.

Fonagy, P., & Bateman, A.W. (2016). Adversity, attachment, and mentalizing. *Comprehensive Psychiatry.* 64: 59–66.

Fragouli, E., & Tourlaki, D. (2020). Leading change and crisis: how transformational changes can be successful nowadays. *Global Research in Higher Education.* 3(3): 22.

Fredrickson, B.L. (2013). *Love 2.0: Creating Happiness and Health in Moments of Connection*. New York: Plume.
Fowkes, M., Fowkes, R., & Solnit, R. (2023). *Dear Earth: Art and Hope in a Time of Crisis*. London: Hayward Gallery Publishing, p. 27.
Gerhardt, S. (2010). *Selfish Society*. London: Simon & Schuster.
Howarth, S. (2025). *Everyday Wonder: How to Find Beauty in the Ordinary*. London: Hoxton Mini Press.
Jean-Berluche, D. (2024). Creative expression and mental health. *Journal of Creativity.* 34(2): 100083.
Keltner, Dacher (2023). *Awe: The New Science of Everyday Wonder and How It Can Transform Your Life*. London: Allen Lane.
Korteling, J.E.H., Paradies, G.L., & Sassen-van Meer, J.P. (2023). Cognitive bias and how to improve sustainable decision making. *Front Psychol.* 28(14): 1129835.
Levinson, H. (1972). Easing the pain for personal loss, 50:5. *Harvard Business Review*, September–October, pp. 80–88.
Lutunatabua, T.Y. (2023). Nothing is inevitable. In Solnit, R., & Lutunatabua, T.Y. (Eds), *Not Too Late: Changing the Climate Story from Despair to Possibility*. Chicago: Haymarket Books.
Macy, J., & Johnstone, C. (2022). *Active Hope: How to Face the Mess We're in with Unexpected Resilience & Creative Power*. Novato, CA: New World Library.
Magsamen S. & Ross I. (2023). *Your Brain on Art*. Edinburgh: Canongate Books.
Markonish, D. (ed) (2022). *Explode Every Day: An Inquiry into the Phenomena of Wonder*. New York: Prestel.
Mason, K. (2014). Becoming Citizen Green: Prefigurative politics, autonomous geographies, and hoping against hope. *Environmental Politics.* 23(1): 140–158.
Mayhew, B. (2006). Between love and aggression: the politics of John Bowlby. *History of the Human Sciences.* 19(4): 19–35.
Music, G. (2014). *The Good Life: Wellbeing and the New Science of Altruism, Selfishness and Immorality*. London: Routledge.
Neff, K.D., & Davidson, O. (2016). Self-compassion: embracing suffering with kindness. In Ivtzan, I., & Lomas, T. (Eds.) *Mindfulness in Positive Psychology* (pp. 37–50). London: Routledge.
Nejade R.M., Grace D., & Bowman L.R. (2022). What is the impact of nature on human health? A scoping review of the literature. *Journal of Global Health.* 12: 04099. https://doi.org/10.7189/jogh.12.04099
Nisa, C.F., Bélanger, J.J., Schumpe, B.M., & Faller, D.G. (2019). Meta-analysis of randomised controlled trials testing behavioural interventions to promote household action on climate change. *Nat Commun.* 10: 4545.
Nisa C.F., Bélanger J.J., Schumpe B.M., & Sasin E.M. (2021). Secure human attachment can promote support for climate change mitigation. *Proc Natl Acad Sci USA.* 118(37).
Nielsen, K.S., Clayton, S., Stern, P.C., Dietz, T., Capstick, S., & Whitmarsh, L. (2021). How psychology can help limit climate change. *American Psychologist.* 76(1): 130–144.
Nielsen, K.S., Cologna, V., Bauer, J.M., Berger, S., Brick, C., Dietz, T., Ulf, J., Hahnet, J., Henn, L., Lange, F., Stern, P.C., & Wolske, K.S. (2024). Realizing the full potential of behavioural science for climate change mitigation. *Nature Climate Change.* 14: 322–330.
Oskis. A. (2019). Not trying to avoid the bridge: avoidant attachment from research to clinical practice. In Cundy L. (Ed.), *Attachment and the Defence against Intimacy: Understanding and Working with Avoidant Attachment, Self-Hatred, and Shame.* (pp. 23–35). London: Karnac.
Parker, M.C. (2023). *The Power of Wonder: The Extraordinary Emotion That Will Change the Way You Live, Learn, and Lead.* London: Hay House.

Prade C., & Saroglou V. (2023). Awe and social conformity: awe promotes the endorsement of social norms and conformity to the majority opinion. *Emotion.* 23:2100–2104.
Piff, P.K., Singhal, I., & Bai, Y. (2025). Bridging me to we: Awe is a conduit to cohesive collectives. *Curr Opin Psychol.* 62, 101979.
Porges S.W. (2022). Polyvagal Theory: A science of safety. *Front Integr Neurosci.* 16: 871227.
Randell, R. (2005). A new climate for psychotherapy? *Psychotherapy and Politics International.* 3: 165–179.
Randell, R. (2009). Loss and climate change: The cost of parallel narratives. *Ecopsychology.* 3: 118–129.
Richardson, M., Passmore, H.-A., Lumber, R., Thomas, R., & Hunt, A. (2021). Moments, not minutes: The nature-wellbeing relationship. *Int. J Wellbeing*, 11: 8–33.
Rosen, M. (2019). *Book of Play: Why Play Really Matters, and 101 Ways to Get More of it in Your Life*. London: Wellcome Collection.
Rust, M.-J. (2020). *Towards an Ecopsychotherapy*. London: Confer Books.
Schmitz, T.W., De Rosa, E., & Anderson, A.K. (2009). Opposing influences of affective state valence on visual cortical encoding. *J Neurosci.* 29(22): 7199–7207.
Servi, L. (2016). *Art is Everywhere: How to Really Look at Things*. Amsterdam: BIS publishers.
Siegel, D.J. (2022). *IntraConnected: MWe (Me + We) as the Integration of Self, Identity, and Belonging*. New York: W.W. Norton.
Simpson, P., & French, R. (2006). Negative capability and the capacity to think in the present moment: some implications for leadership practice. *Leadership.* 2: 245–255.
Solnit, R., & Young-Lutunatabua, T. (Eds) (2023). *Not Too Late: Changing the Climate Story from Despair to Possibility*. London: Haymarket.
Stancato D.M., & Keltner D. (2021). Awe, ideological conviction, and perceptions of ideological opponents. *Emotion.* 21: 61–72.
Weintrobe, S. (2021). *Psychological Roots of the Climate Crisis: Neoliberal Exceptionalism and the Culture of Uncare*. London: Bloomsbury.
Weintrobe, S. (2013). *Engaging with Climate Change: Psychoanalytic and Interdisciplinary Perspectives*. London: Routledge.
Whybrow, A., Turner, E., Mclean, J., & Hawkins, P. (Eds) (2023). *Ecological and Climate-Conscious Coaching: A Companion Guide to Evolving Coaching Practice*. London: Routledge.
Winnicott, D.W. (1971). *Playing and Reality*. London: Routledge.
Zuabi, A.N. (2023). The Herds, a vast act of theatre to spark climate action. *TED Conferences [Video]*. Available at: ted.com/talks/amir_nizar_zuabi_the_herds_a_vast_act_of_theater_to_spark_climate_action

Interdisciplinary resources

Action for Happiness: A registered charity offering coaching, talks, courses and local volunteer-led groups. actionforhappiness.org/groups
Artists and Climate Change: artistsandclimatechange.com
Center for Compassionate Leadership: centerforcompassionateleadership.org
Climate Psychology Alliance: Supporting individuals and groups struggling with eco-distress and offering safe spaces to share emotions surrounding the climate crisis. climatepsychologyalliance.org
Climate Stories Project: climatestoriesproject.org
Creative Carbon Scotland: cultureforclimate.scot/services/arts-into-climate/
Deb Dana: rhythmofregulation.com

Extinction Rebellion Families: extinctionrebellion.uk/act-now/local-groups/#families
Extinction Rebellion Art Group: extinctionrebellion.uk/act-now/resources/art-group/
Forestry England Arts: forestryengland.uk/arts
Greater Manchester Environmental Trust: gmet.org.uk
Imagination Institute: Led by psychologists Martin Seligman and Scott Barry Kaufman, the USA-based institute is dedicated to making progress on the measurement, growth, and improvement of imagination across all sectors of society: imagination-institute.org
Invisible Dust: UK-based Invisible Dust creates new thinking to address urgent climate futures through ambitious collaborations between leading artists and scientists. Invisibledust.com
Julie's Bicycle: Julie's Bicycle is a pioneering not-for-profit, mobilising the arts and culture to take action on the climate, nature and justice crisis. Their "Creative Green" programme works with individual organisations and networks through consultancy and partnership. juliesbicycle.com
London Wildlife Trust: wildlondon.org.uk
Metal: metalculture.com
Moral Imaginations: UK-based research and practice centre dedicated to training "Imagination Activists" – a new kind of activism powered by imagination to create new possibilities and forms of action. Moralimaginations.com
Murmur Earth: murmur.earth
Not Too Late: Solnit & Lutunatabua offer a practical guide to what they can do against climate chaos and for a just and thriving natural and human world. nottoolateclimate.com
Outdoor Arts UK: outdoorartsuk.org
Radiate Arts: climate.cymru
SALT + EARTH: creativefolkestone.org.uk
The Natural History Consortium: bnhc.org.uk
The Herds: theherds.org/get-involved/climate-actions
University of the Arts London's Climate Emergency Network: arts.ac.uk/about-ual/climate-action-plan/climate-emergency-network
Wild Rumpus: wildrumpus.org.uk

Chapter 6

Uprooted

Land, loss and new shoots

Linda Cundy, Myriam Sarens, Rebecca Attoe, Dawn Whittaker and clients at the Grounding Project

This chapter considers attachment to place, and specifically to natural environments that are personally significant. With the input of staff and members of the Grounding Project, a horticultural therapy project for refugees in the London borough of Southwark, it explores the influence of the landscape and ecosystems we grow up in on attachment and identity. Members of the project share their stories of early life, their engagement – or not – with the natural world during their formative years, and their experiences of being in exile from the environment of their childhood. They also describe the healing potential of connecting, or reconnecting, with the natural world of plants, soil, birds, insects, water and light, and the seasonal cycles through gardening. "Roots, I think. It's always about roots. They connect us to something intangible – our ego and identity – and they connect us to our home" (Snow, 2019, p. 300) – roots in both senses of the word.

While those clients who contribute to this chapter sought refuge from war or persecution, the plight of those displaced by climate crises is also considered, and the likelihood that a significant number of British citizens will be internally displaced due to the impact of global warming in the near future.

***

Context: Ancestral roots

How many generations have passed since your forebears had a close connection to the land? For all of us, those years would be a mere moment in evolutionary terms. We human beings are the most adaptable of all animals and can live in a concrete jungle of asphalt roads, high-rise buildings, concrete, glass and steel, yet our phylogenetic relationship with the soil, trees, plants and other creatures remains like a powerful echo.

In June 2024, to celebrate Refugee Week, I attended an event organised by CTAR (Essex University's Centre for Trauma, Asylum and Refugees). The theme was "home," and numerous speakers from around the world shared stories about the meaning of home for those who are exiled from theirs. Of the

DOI: 10.4324/9781003733942-6

many heart-rending and uplifting examples, one made the most powerful impact; a Palestinian man driven from his home and land in the occupied West Bank decades before, still living in a house he considered temporary. While some spoke of taking seeds and cuttings from their gardens when they were forcibly evicted, he also took a bag of soil. This moved me deeply and set me reflecting on themes of origins, identity and attachment to place; to slightly misquote Wendell Berry, *the earth we are shaped in*.

We were all creatures of the land till very recently, in terms of the evolution of our species. There are few hunter-gather societies that still exist, but until around 12,000 years ago all *Homo sapiens* were just that, following game and other sources of food through the seasonal cycles and migrations. Then, in the Fertile Crescent that links the Persian Gulf with the Mediterranean, agricultural life began by domesticating cattle and other creatures, and sowing and harvesting crops. Our relationship with nature began to change as we started to "own" it, put down roots, to keep and breed creatures and propagate plant species for our own purposes. Once most of our ancestors had settled, specialist occupations developed and urban life started with all the necessary infrastructures. Our sense of identity no doubt shifted too, as we domesticated ourselves (Roberts, 2017). In Europe and the Global North, before the end of the 18th century, almost everyone still owned, worked on, or had access to some land. The Industrial Revolution changed that for many. In most other regions of the world, however, involvement in agriculture is ongoing.

Bowlby was influenced as much by Darwin's theories as by those of Freud, viewing human nature as shaped by both natural and social influences. Each infant adapts to its relational and cultural ecosystem, but also to its physical environment. Attachment is a survival strategy, and we each learn from our attachment figures about the dangers present in our immediate corner of the world and how to keep ourselves safe from predators, venomous creatures, those that carry potentially deadly diseases, climate extremes and so on. We also learn other survival skills, including how to feed ourselves. For those millions who are now alienated from the soil, who have no space or inclination to grow food, that may simply be a visit to the supermarket or takeaway. But still, for so many around the world whose lives remain embedded in the land, knowing what is edible or poisonous, how to farm livestock and practice agriculture in a manner suitable for the specific climate, is essential. In Papua New Guinea, small Indigenous children of around 5 or 6 wander barefoot through forests hunting huge venomous spiders. They lure them from their lairs with a stick, grab them skilfully to avoid being bitten, then make a fire to bake them for food. And in 2023, a plane crash in the Colombian Amazon made headlines around the world. Four children from the Huitoto Indigenous tribe had been on board with their mother. All the adults were killed, but the children, aged between just 1 and 13, survived alone for 40 days because they knew the environment intimately, its dangers and resources.

For those of us in the developed world, our relationship with nature, the other-than-human, is very different but still survives. "Nature inserts itself into the mind of man" (Holmes, Chapter 1 in this volume), but it is also deeply embedded in our emotions, imaginations, even our physiology. Just as tree rings indicate the years of plenty and hardship, in terms of nutrients, water and light, so our genes, brains and nervous systems register nutritional and relational times of hardship and trauma, deprivation or abundance. Millions in the industrialised north adapted to an urban life working in mills and factories, and the mining industry that provided coal and other minerals to keep them fuelled took many more underground. At that same time, Romanticism was flourishing in the arts with such painters as Constable, Turner and Caspar David Friedrich, authors including Thomas Hardy, and poets Byron, Keats, Wordsworth and Blake (who, in *Milton: A Poem*, wrote of the "dark satanic mills" at the start of the 19th century).

***

Not only the industrialisation and urbanisation of societies but also famine, war and greed have led – and still lead – to people permanently expelled from their homes, including the Palestinian man whose bag of earth from his garden inspired this chapter. The ache and rage of injustice are difficult to come to terms with.

A client of mine, Beth, reconnected with her intergenerational narrative during our work together, having "forgotten" that her maternal grandmother was of dual heritage. Beth's great-grandmother was the daughter of an American ranch owner, while her father (Beth's great-grandfather) had been a Cherokee ranch hand. He was banished when the child was conceived, but he later kidnapped her and took her to live with his people on a reservation. Feeling moved and inspired to discover this part of her own history, of *herself*, Beth visited a reservation, met tribal elders and learned about their history and culture. The Cherokee, along with other Native American peoples, had been highly skilled farmers with a unique approach to growing crops. The "three sisters," their staple vegetables, were corn, various types of squash and climbing beans, which were planted together; first the corn, then the beans that climbed the maize plants, with finally the squash seeds sown in between. This method kept the land fertile – and their cultivated fertile soil was coveted by the "white man." The final blow for them was the discovery of gold in their ancestral lands. This led to the Indian Removal Act in 1830 and the compulsory eviction of the Cherokee, along with other Indigenous peoples, over vast distances to designated "Indian Territories" in the West. During this forced march, the Trail of Tears, an estimated 4,000 Cherokee died. The allocated land was unsuitable for agriculture. Without soil to farm, alienated from the landscape of their peoples, depression and addictions spread like wildfires. These productive people were no longer embedded in their natural environment, lost purpose and were at risk of losing their identity.

When Beth learned about this historical trauma and injustice, it helped her make sense of many aspects of her own life and identity; her grandmother's depression passed down from mother to daughter and beyond. While visiting a Cherokee reservation, an elder told her, "It takes seven generations to heal a trauma." Seven generations for the rage and grief to subside, for a people to redefine their sense of self, for an updated culture to emerge, and perhaps for feeling truly at home. Doing the maths, Beth reckoned that she was the seventh generation of her First Nation family since the Trail of Tears.

Of course, similar outrages and atrocities have happened the world over throughout history. With the industrial scale of slavery, again mostly from the late 18th to mid 19th centuries, those taken from their African homes were destined for enforced labour growing cotton, sugar and other crops in landscapes and climates very different from their own. In the United States, once slavery was abolished, many people of African heritage stayed on the land and became smallholders. Out of almost a million black farmers in the United States in 1920, "a staggering ninety-eight percent were dispossessed of 12 million acres of land" by "theft," federally orchestrated and "backed by para-military violence" by the end of the 20th century (Táíwò, 2025, p. 53). Land ownership is power, and yet again it was the disenfranchised who were cast out.

***

Ingredients of identity

Achotegui writes: "identity is constituted by all of the fragmentary introjects of childhood. This process takes place in the framework of relationships with one's parents, families, and community." But I believe that the physical, ecological context in which we develop is also introjected and plays a significant part in one's identity. There are environments and ecosystems that are meaningful to us. I think back to the landscapes of my early childhood, places I explored, played, observed, and where I learned about the natural world. These were the backdrops to the rhythms of daily life. My father served in the British army so it seemed we were uprooted whenever I developed a strong attachment to place, but each feels to me to be a core ingredient in the person I became; red squirrels and fallow deer in the dark forests of Germany (so poetically captured in the paintings of Franz Marc that hung on every classroom wall during my childhood); picking field mushrooms and blackberries with my father, watching rabbits in the evening light, the tinkling of crystal-clear streams, rockpools and the strong aroma of seaweed in Dorset. Then the harsh light, searing heat, dusty red sands, lizards, scorpions and snakes, camels and goats, tumbleweed, cacti and tiny frogs after the rains in Libya. I recall the ache of leaving each of these, the grief that felt too adult an emotion for a young child. While relationships with people were important, it is the sensual memories of place that I most easily conjure up.

Many of us have travelled to distant countries to experience the scenery, flora and fauna. (I now baulk at the thought of all the air miles I have flown and offset what I can with carbon capture and mangrove planting.) We may try to recreate the exotic tropical, Mediterranean or Japanese gardens in our own back yards. Others choose to move to very different natural environments for study, work, or love, knowing they can return to their homelands when and if they choose, and be richer for their experiences. It is one thing to choose to leave, knowing you can return. It is a different matter to be forced from one's home.

In a recent online news article, BBC journalist Lucy Williamson reports on scenes witnessed by herself and her photographer of an attack on a Palestinian olive farm in the West Bank by Israeli settlers. Photographs show hooded men racing towards farmer Brahim Hamaiel's property and setting fire to homes and vehicles. They have already hacked down his olive trees. Williamson reports that this intimidation is a frequent occurrence:

> "Fear is natural," Brahim had told me, looking up at the ridge where tarpaulin flapped at the settlers' lookout post in front of a few caravans and makeshift homes. "But there's something stronger than fear that drives me to stay here – the scent of my ancestors and an attachment dating back hundreds of years – even if I pay the price with my blood."

As Williamson (2025) notes, "In the language of this conflict, each birth and each burial only strengthens the ties to the land."

But what is it like to lose the natural environment in which our identities took shape? To be exiled?

***

The Grounding Project

The Grounding Project is a community-based South London and Maudsley (SLaM) NHS horticultural therapy service supporting refugees and asylum seekers who have endured traumatic life events in their countries of origin – war, torture, imprisonment, sexual violence, persecution, trafficking and slavery. All are suffering from complex PTSD, often compounded by anxiety disorders, depression and suicidality. Around 80 per cent also have some degree of physical disability as a result of their experiences.

It is open for one day per week, offering trauma-informed care for those who struggle with mainstream services. The garden allows individuals to connect with the natural world, build a community, and receive support from qualified clinical practitioners. According to Sue Stuart-Smith, "there are many ways in which anger, grief and frustration can be sublimated or channelled creatively and gardening is one of them" (Stuart-Smith, 2020, p. 42). If a client is hyper-aroused, or hypo-aroused and detached, going into a fight, flight or freeze state, there are activities in the garden that can help with self-regulation; discharging energy through more intense physical activity such as digging; the quiet work of sowing seeds and potting up cuttings to calm the nervous system; gentle grounding by feeling the soil, engaging all five senses to help dissociated clients transition back into their bodies. There is also a trauma-informed yoga group for those who choose to attend it.

Since 2017, the project has been based at Roots and Shoots, an educational and environmental charity in Kennington, London. This partnership provides stability and continuity for clients, including a dedicated garden where participants grow fruit and vegetables in a safe, supportive environment. There is a large greenhouse, raised beds for planting, the Calming Garden, and a wild garden.

I have visited the Project on a number of occasions, including their Open Day event during World Refugee Week 2025, meeting client members who very kindly agreed to be interviewed for this chapter, and staff who collaborated with me.

Most of the clients who took part spoke about their early lives in their home countries, but not specifically about trauma or their reasons for having to leave. They also reflected on the significance, for each of them, of the horticultural therapy provided at the Grounding Project.[1]

***

Roots: early landscapes

1 B.

> Okay. I am B … I came from East Africa, my country is Somalia. I've been here about 27 years. Back home we have a farm – we live in the countryside. I grew up there, but at the time I was young.

B spoke of the livestock – cows, sheep and camels – they kept, as well as cats and a special dog, Abla, who protected the sheep from hyenas and other predators. There were also many monkeys that had to be chased away. Near the farmhouse was a big old tree, a famous local landmark, that provided shade.

> I wanted to live there and stay there. But the civil war started, so we have to run away. We left behind everything.

B's family returned to their farm when the civil war ended in Somalia; "Still the land there. So whoever they left in the family, they go back to that. We still got the land." But B had experienced so much life-threatening horror that she could not go back. However, her family send her photographs which she showed to me on her mobile phone, the sheep, cattle and camels grazing on parched land in a spectacular landscape of hills rising to mountains. One photograph showed writing in the sand: "They write my name and they take a picture and send it to me. They are in the sun and that's back home. Here the sun is weak, so they send this to me."

2 L.

> Back home in Kenya, we used to have a farm. Every weekend and holiday if we're not going to school we are at the farm. We planted potatoes, maize, kale, cabbage. Everybody used to go and farm their portion. And when we finished, our mum used to ask us to help the neighbours. If you don't finish, you farm the next day until everybody finished.

She described the tea plantation in detail – picking the fresh tips from the bushes, putting them into a basket on her back, carrying the heavy load up to the road to be collected by lorry and taken to the factory. But with the coffee beans,

> We used to take it home, and then we dried it and make coffee. And we brew some coffee and maybe we give the neighbours. My mum used to say, take it to the elderly people, the needy people. We never used to sell it – we used to give it to them and then leave some for ourselves.

Also on this fertile, productive farm they kept cattle, chickens and ducks, and grew fruit, including passionfruit.

So many of her memories included her community of neighbours, working together, looking out for each other, sharing produce; the land and the people very much one and the same in her inner world.

3 Y.

Unlike L, who lived and went to school in the village near the family farm, Y grew up in Ethiopia's capital city, Addis Ababa. His grandmother owned a farm 45 km away.

> My grandmother, she had a huge farmland, and she had animals like cows, ox, chicken. We plant different kinds of trees, fruits. Some of them are not here, only in Africa. For example, I like mango very much. Banana also. You know, my grandma she have a lot of bananas. We have our own avocados too. And we harvest a lot and we even sell to the community, the local area. Vegetables like onions, beets, different kinds of cabbage. We have flowers as well and sell a lot to people from the area for weddings, graduations, and so on. We are a general supplier. We employ a lot; we pay them.

Y. was sent to boarding school in Addis Ababa. During the vacations, especially the long winter holiday in July and August,

> I'm not allowed to go with my grandma because my dad, he doesn't want to go there ... Always he scared me – "if you do something wrong, I'll send you there for punishment." So I say no, sorry dad, I'm not going there, I don't like it ... But I will do some mistake deliberately, and my dad sent me for punishment!

Y described helping on the farm during those months of "punishment." His grandmother taught him how to plant garlic, onions and beans in straight rows. He helped with general farm jobs and also learned about beekeeping.

> When the honey is ready, you know, we harvest like in the traditional way. We make a lot of smoke – they don't like smoke. We don't have gauze or something like that for protection in Africa. They just put in a lot of smoke, and the bees leave the hive, and they can get to the honey.

However, on one occasion, things went wrong, and he was badly stung: "My face, my eye, everything is big, big."

4 O

O, originally from Guinea, also grew up in a city centre, but visited his family in a village during holidays. His grandmother grew cassava, rice and fonio (*Digitaria exilis*, an "ancient grain" that is more drought resistant than most other staple crops), and he enjoyed helping her with cultivation.

5 N

N recalled with passion his early life.

> I'm from Bosnia, so the only thing we bought, which I remember, it was like oil, washing up powder, sugar, because we could produce the rest. We had our own milk, meat, eggs. We would make butter, we would make cheese, I mean everything. Every house had chickens, cows, horses ... plum tree, walnuts. apple tree, pear, so every single house we would make juice ... We've been there for thousands of years. Look, each generation you learn everything from your father, grandfather.

He described various methods he had learned for preserving foods to keep through harsh winters. Although now in London, he has only a balcony for outdoor space; he does grow some food crops and teaches his children about gardening and cooking, passing down the horticultural knowledge of his ancestors.

> It's beautiful world where everything is beautiful, why, why … I mean, look at some people you know who maybe never touch seed, plant the seed, put little bit water, easy to understand. And brother you have food, you have food, you survive. That's why I, you know, trying to teach my kids. Soil is not dirty, soil is, I mean, compost is maybe much healthier than stuff you buy in the supermarket … We need to go back to simple things, you know, to enjoy life.

***

According to Mary-Jayne Rust, "our relationships with the more-than-human world are woven into, and inseparable from, relationships with self, family, culture and the earth" (Rust, 2020, p. 89). For those who spoke about the environment of their childhood, I had a sense that they were reconnecting with something important and vital for their identities, and reanimating the memories of significant attachment figures who live on in their inner worlds. Being taught to fertilise and care for soil, sow seeds, nurture young plants, harvest crops and turn them into meals by the side of a trusted adult creates a special bond. Y's grandmother, who owned and ran a huge successful farm, made time to pass skills on to him, as did O's grandmother, teaching him to plant rice, cassava and other crops, while N learned horticultural expertise passed down for generations through his male lineage. When B recalled and spoke about Abla, the sheepdog, a warm smile lit her previously serious face.

> Abla used to look after the sheep, look after the house. She put away all the stranger people. So, you know, because we have like hyena that come in the night and eat the sheep. So she looked after them well. I'll never forget that.

It seems that this precious dog represents safety and protection, a non-human secure attachment that continues to exist as a "secure internal object." Yet, she reminded me, "In my country people do not like dogs; they think they are dirty creatures."

For N, it is the fertile land of his early life itself that seems to run like a seam all through him, the core of his personal philosophy and identity. "*You know, we, nature, are one*."

Despite the traumatic events that shattered the trajectory of these people's lives, reconnecting with their memories, both through speaking with me and through the embodied activity of planting, potting, staking, feeding and watering plants and harvesting produce, helps to preserve the sense of something good and healthy, untouched by torture or war.

Uprooted – and new shoots of recovery

Most refugees are in exile, unable to go back to their previous lives and possibly risking imprisonment, torture or death if they were to return. For some, traumatic memories are too hard to erase, and sights, smells, even language can trigger a trauma response. Memories of happier times may be easier to tap into from the – relative – safety of distant shores. Yet their homeland still exists; regimes may come and go, but the land itself remains. Homes and farmland may now be occupied by strangers, perhaps by people responsible for driving the original inhabitants away. War may have destroyed much of what was familiar. But still, something of the landscape survives. This can create a painful dilemma, an irresolvable loss, migratory mourning (Achotegui, 2019). I have met a number of refugees and asylum seekers who suffer deeply from the realisation that they cannot be buried in their motherland, beside their family and in the earth that shaped them. This is not about a country as a political entity but a profound relationship with its soil.

***

Some members of the project felt unable to talk about their childhood homes, either because of the trauma associated with that life or because the pain of loss was too great. However, they were keen to share their experiences of horticultural therapy at the Grounding Project.

B, from Somalia, spoke of it as her "Monday medicine":

B: I have a condition – fibromyalgia – that make me down, moody. Sometimes you cry, cry, cry, then it comes to my mind, "Oh, I'm going on Monday. So how far is Monday?" And feeling better. You know, happy, mood change. I cannot explain how it bring back life. You have people have same situation as you. Sometimes I feel like to stay here. Sleep here.

Linda: You would like to live here, at the Grounding Project?

B: Yes. Here I feel safe. Normally, I don't trust people [because of what] happened to me in the past. Here, I've had a chance, people have talked to me. But these people, I'm thinking that I have to stick with, hold them there forever. I talk from my heart.

This contrasts with her experiences within the local Somali community: "If I share my problem with my community, it kind of like, it's your fault something happened to you. Everybody talks behind you, in front of you they're calling you name. If you have mental problem you're crazy."

She spoke with deep appreciation of the support she received during the Coronavirus lockdowns, when it was not possible to attend the Project or meet up. The weekly phone calls and other contacts helped her cope with the isolation and hold onto hope.

While the community takes centre stage for B, the sense of peace created in the green space surrounded by the bustling, alienating cityscape – so utterly different from the huge empty space of desert and mountains of her family's land – is an oasis of safety where these relationships can unfold and be sustained. I am reminded of the great tree outside the farmhouse back home, a landmark providing shelter.

***

L from Kenya:

> [Being here at the Grounding Project] – it's reminding me of our soul. I love nature and how we plant these seeds and see them growing every week. It's good, and it's calming around the garden. When I come here [to the UK], I don't feel like I'm ... London is noisy and grey and you know in here it's peaceful.

Like B, the sense of community is equally important for L, the neighbourly work of planting seeds together, watering, nurturing, harvesting, helping and sharing, just as she learned to do in childhood. I imagine this connects her to that inner core of good, nurturing memories, of early safe, supportive relationships and values, before trauma disrupted that peaceful, productive life.

***

N from Bosnia:

> I was involved in a war. So came here with this and then, for maybe ten, 15 years – I don't know, I noticed that I was changing but I didn't know what happened ... So I had like a mental breakdown then ... GP sent me to the Maudsley Hospital and it was explained to me what's happening with me ... [Here at the Grounding Project] I'm doing this gardening stuff and producing, you know, plant from seed. I mean, it was a revelation. It was something much better than tablets or injection or whatever, talk. It just ... brings me back to happy, maybe before when I was younger, growing, you know, before that war thing. Happy memories.
>
> I mean, this project, it's unbelievable. For me, it's really good, it's ten out of ten.

Although many of the Project's members grew up on farms, the climate, seasons, and terrain were very different from those of northern Europe, so the crops they were familiar with are not grown here. For N, most of the crops in Bosnia are also found here in the UK. He is able to share his skills and knowledge about these plants with others in a very diverse group. However, he does appear to have a favourite: "They call me Tomato! So they [tomato plants] are like my kids. It's everything, because I grow them at home [in Bosnia] ... We have to live in nature. You know, big cities, everything in plastic ... You know, we, nature, are one."

***

O, from Guinea, told me,

> Before, I used to sit at home all the time. I felt depressed, sad, negative, but now I'm happy to come. It's helped me to talk, to meet people. I like gardening and that helps me too. It allows me to keep positive. Watering, planting, I like everything.

Y from Ethiopia:

> I was persecuted in my country. I lost everything. My grandma, my dad ... Here, I cannot even travel because I don't have any status, I don't have any papers. You know, I have a lot of struggle in this country.

Y was referred to the project by his NHS psychotherapist when she heard about his love of gardening. "This is helping me. Every week I come here – Monday is my special day. When we harvest some tomatoes here, spring onions, it's reminding me of back home."

As with the others, it was the combination of being in a green, safe place, engaging in gardening, and also the supportive community that Y finds so healing.

> Because I do not have any family here, the only family I have here is this group, like brothers and sisters. We have different nationalities, language, culture. Many experiences also, but we are one here. We have one purpose: to restore our life. To forget about our trauma, our scars, our stigma.
>
> ...
>
> Nobody helped us except Myriam and her co-workers. I would like to say thank you to them because they support us a lot. Without them, we are not alive. Our life is better than before.

Y was keen to also thank the doctors, nurses, psychotherapists and other staff at the Maudsley NHS Trust who have helped him so much in his recovery from trauma.

A further factor in recovery from the mental and emotional scars for Y is his concern to help others. "[I want to be] a voice for the voiceless, for my life experience to help others. That's why I wrote my poem."

At the Calm Garden

We can live in harmony
With our natural environment.
Colours of green from a cool palette,
Presenting a calm spirit of the city.
The opportunity to enjoy
Its unspoilt nature,
To provide a heaven for relaxation.
A place to strengthen
Mental health and wellbeing.
The journey to an enchanted,
Life-changing sanctuary.
The healing power of nature
At the Calm Garden, at Roots and Shoots.

A sanctuary for survivors,
A safe space for anyone going through
Their journey of recovery
Making their mental health a priority.
A welcoming place for those seeking safety.
For migrants and asylum seekers
It creates a culture of welcoming …
Belonging is beneficial to everyone.
Gardening is our therapy and
Makes a positive change for sanctuary-seekers like us.
At the Calm Garden, at Roots and Shoots.

We are dreamers,
Believers and creatives.
This lifts our spirits and hopes,
Heals our minds and hearts.
A place of sanctuary and peace.
Fill it with sunshine.
The rain will fall,
The wind is gently blowing.
Spring is here!
Everything is growing.
We take care of our garden,
We harvest joy: a reward so sweet
At the Calm Garden, at Roots and Shoots.

A true sanctuary for survivors.

(Poem by Y, 2024)

***

Later, I met with K in the wild garden, away from the greenhouses and vegetable beds, under a big oak tree. Unlike many of the others I spoke with, K had little contact with nature during his early life. Coming from Iraq, his father's work meant the family travelled around, staying a year or so in one city, then moving on to another and another. He now has numerous trauma-related symptoms, including problems with his memory.

> I had a psychologist which was face-to-face for a couple of years. I was just struggling at home. I didn't go out, just go to the meeting and then I go home and the doctor was suggesting to go to – we have a place to go for gardening. Just try it and check how it affects your mental health. It might be you get benefit from this and I said, okay, let me try. And when I met them, it was okay for me, and when they teaching us to plant the seeds and put the plants in the ground and growing this one it's making me less stressed during the daytime. I love it, and I feel more freedom and less stress.
>
> I told them I'd like to have minimum two days a week, and they say we can't afford it. One day is not enough for our people because our mental issue is not stable. I have a post-traumatic stress disorder, and sometimes I totally change, different person. This is out of my control.

Although he appreciated seeing the same faces at the Grounding Project each week, K was drawn more to private contemplation than company. In this green oasis surrounded by tall buildings and bustling London streets, he can find some peace.

> I like to walk around here and see the nature in this wild garden. I like to touch the leaves because I feel numb so many times in my body. And when I touch, I feel there's something sometimes … Sometimes I … I lost all my sensitivity. When I walk, I feel I'm walking on the sea and all my body's numbness. And I have to touch something. My psychologist teach me to touch and to smell leaves. That's what I try to do.

K slipped off his sandals and wriggled his bare feet in the thick layer of leaf mould under the oak tree. I had a sense of him taking root in the earth there in the dappled late summer light … one place he feels safe.

> Feet on the ground. That's it. Like I move my feet – I'm actually different, you know, I feel it's different. More alive. I just like watching the plants and the butterflies, everything. I don't mind sitting two hours, three hours, just watching the trees. Just watching some small plant for me is fine. I *love* it. There's so many leaves. They're so different, so different. Amazing.

Notably, for K it was not so much the horticultural therapy – the gardening – that is healing but simply being immersed in nature, breathing it in, being part of it. Nature is his safe haven.

Whether they had owned land and animals back home that had to be abandoned or, like K, had been urban dwellers, in this small community garden, a green and fertile island in a sea of tarmac, concrete, glass and steel, these refugees are finding a new landscape to internalise, a place to belong, to find meaning, community, and recovery from trauma.

***

After, I had a conversation with staff members Myriam, Rebecca and Dawn about the work they do.

Linda: Can you tell me about the Grounding Project? I know there are different ways of running horticultural therapy services. For instance, I think at Freedom from Torture the psychotherapist works alongside a gardener as part of the horticultural therapy.[2] But you do things differently here, so I'm interested in the philosophy behind that.

M: The philosophy here has really grown organically in response to our client group's needs. We want this to be a sanctuary for people. This is a therapeutic space, where people can find connection to the natural world and to each other through horticultural therapy. Most of our clients have experienced a lot of powerlessness, and we're giving choices by working in collaboration and encouraging clients to take the lead so they can regain a sense of agency.

We're not doing trauma-focused therapy work here – that tends to take place in individual therapy sessions. We keep this as a very safe space for people where they're not going to be triggered by each other's stories. There is the power of the unspoken, and an understanding that they have all suffered – this creates a powerful bond and a sense of connection. But we think a lot about trauma-informed work and the garden project functions that way, giving people choices in how they interact in the space and with each other so they can regain a sense of agency, empowerment and community.

We see how nature helps clients in regulating and calming their nervous system by engaging through the senses in the present moment. It is also a culturally sensitive intervention that many clients relate to more than talking therapy, which is often unknown or stigmatised in their culture.

D: I think the fact that we don't talk about people's individual trauma in the group is really important, actually. And there's a huge respect for that, because they all know that they've all been through terrible things. I'm sure they individually do talk to people, you know, they become friends with each other, and I'm sure they talk. But there's a deep understanding of each other's pain, I would say, and the respect not to intrude on that.

R: We try to keep it as inclusive as possible. Not everyone speaks fluent English, so the design of our project means that everyone can still engage at whatever level they're comfortable and get what they need. It doesn't rely so much on language skills. Also, not everyone in our group is able-bodied, so we try to make it as accessible as possible for them. All our growing beds are raised, designed to be very accessible. For those who, for physical or emotional reasons, who can't be part of a group, we find other things, and that's what makes it so safe.

M: The beauty of horticultural therapy is that people can engage in their own way. Some want to do something very quietly or with one person; they might want to sow some seeds in the greenhouse or just have their own space. Others like to chat and pick vegetables, water plants and be quite active, which is also possible.

R: It doesn't matter whether you're having your best day or your worst, you can come and engage in a way that feels suitable for you, and everybody understands that – they all know those feelings.

D: I think I've realised just how unwell people are, which is much more than I expected. There's an up and down with PTSD. There's not just a healing process when they get better and better. It isn't like that. My sense is that most of the group have not wanted, not felt able to make use of a talking therapy, which is why they're referred to Roots and Shoots in the first place.

R: Watching things grow from seed to fruit is really important for people. Everyone describes it as very hopeful, and you often see they'll write their names on sticks and put it next to a plant. It's like pride in what they're doing.

Linda: Perhaps they see something of themselves, beginning to grow roots and put out new shoots?

R: Yes, being part of that growth is very helpful for people, I think. They resonate with it a lot. Also, we give people a choice about what they would like to grow, and often they pick things from their home country that they want to try here. We always try, give it a go. Everyone understands it might not work, but we can just try and if it doesn't grow, we try something else. I think that's a very important message for them too.

Linda: Some have real growing expertise here, too. I imagine most people feel pretty powerless in a strange country where perhaps they're not fluent in the language, and let's face it, they don't always get the warmest reception. But when they're here, they've got a certain status.

M: They often have a great deal of expertise, especially about growing particular vegetables they have cultivated back home. So we are often learning from them.

R: Yes, some people really are very expert gardeners. We want them to feel they have agency and they can share their skills with everybody. And we do all the jobs and stuff with them, so no one feels othered. We're all on the same level.

D: But then there are other people who've never grown stuff before. And they were just absolutely blown away by growing, you know, from week to week – something's tiny, and the next week it's 10 inches long. And you're thinking, how on earth did that happen?! And growing from a flower to a fruit to something you eat – it is beautiful.

M: We've also had some wonderful opportunities to collaborate with artists here at Roots and Shoots (R&S), and this has resulted in some amazing art projects. The team here have worked with us by essentially developing themes based on what the clients proposed. One such project was "Flock," with a ceramics artist, where clients made birds representing migration and a sense of freedom. This was exhibited at the Horniman Museum. This year, clients made stunning mosaic pots for the garden with help from the artists at R&S. Another was with the Butterfly Conservation Project; they were here for several months, teaching us all about different aspects of butterflies and developing a beautiful butterfly sculpture for the garden. These projects give everyone a real sense of achievement and belonging.

R: That's the good thing about our projects, people just make meaning with it. We were learning about how some butterflies migrate and come from different countries, then everyone in their own way is resonating about journeys. You don't need to be explicit. People pick up the messages about the hope that goes with growing things.

Linda: Did you know that the ancient Greek word for butterfly is psyche? And something else you may have learned is that when a caterpillar is inside its cocoon, it completely dissolves. Then the molecules reorganise themselves into a new form, the butterfly. So it's a perfect metaphor, really, for therapy. I think that you go through this process of kind of going to pieces, dissolving, then you reorganise the narrative, the elements of your identity to create something new. The Grounding Project is like a cocoon where people go through a transformation. They are becoming unique, beautiful butterflies.

R: Wow …

M: But the sense of community here is really the biggest thing, isn't it? People really form such great friendships here, and there's amazing peer support. Refugees often had good early attachments, even though later they experienced ruptured or severed relationships through trauma and displacement.

Linda: Yes, I noticed how often people speak warmly about their parents, grandparents and other close family.

M: This may be why they are so good at forming positive bonds and making new attachments in the group – fostering a caring community reflects positive experiences in early life. So if anyone has problems travelling, for instance, they phone someone up and arrange to come together. There have always been those sorts of things going on. People really care about each other, support each other.

R: Letters, paperwork, appointments. Someone's been through it, and they'll explain it and help others. The community is so important here. That's what makes it work.

Linda: This makes such good economic sense, too, to draw on each other's experiences rather than need to turn to external services. It's so cost-effective. Clearly, the community is an absolute lifesaver for most of the people here. But what about the land, the soil, the growing?

D: I think that's absolutely at the heart of it. Some members will say, I need to just plant something today or get my hands into the soil. And there's quite a lot of joking around gloves – do you wear gloves? Don't you wear gloves? There are certain people here who will never wear gloves because they just want to get their hands dirty.

Linda: I bet N is one of those!

D: Yeah, definitely! Gardening settles them down; it calms them. It brings them into the present moment. You know, it's just so mindful, isn't it, gardening? You can feel the change in mood when we go into the garden. We can be in that room, and everyone's a bit grumpy, and the world is a horrible place, and people are struggling with all their different difficulties, quite angry sometimes about stuff. And then as soon as we get into the garden, everyone's just completely transformed. It's very noticeable. And if the conversations are getting very heated, we'll go into the garden, and it really helps. It's regulating.

***

Linda: Can you say something about how clients come to this project?

M: The Grounding Project is one of the options open to asylum seekers and refugees with PTSD. Generally, the GP will refer clients to IAPT services, they assess and then refer on to us as part of the psychology and psychotherapy pathway for clients who need more long-term and in-depth therapeutic work. Some clients will also choose to engage in individual trauma-focused therapy, such as EMDR and narrative exposure therapy (NET), this is at one of our clinics as part of the trauma pathway.

R: There are psychoeducation groups, too.

M: Yes, before people come to the group they're generally offered psychoeducation, which basically means helping them with tools to ground and stabilise themselves because they all have PTSD, and then they suffer from overwhelming symptoms that they often don't understand. Some people have said they thought they were going mad, and it helps so much to make sense of why they feel that way.

Our service is underpinned by Judith Herman's Phased Trauma and Recovery approach.[3] Clients begin with Phase 1, individual and/or group stabilisation sessions to understand their current difficulties and learn tools to manage their symptoms. After stabilisation, they're evaluated for additional Phase 2 trauma-focused therapy to process traumatic memories, and are invited to join our project as Phase 3, reintegration and rehabilitation, though some opt to come straight to the Grounding Project after stabilisation.

R: When it comes to winter, when perhaps there's not so much growing as we do during the rest of the year, we'll do things like psychoeducation workshops here. We ask the group, is there something you'd like more information about? For instance, it might be sleeping problems, and we'll do a session talking about sleep and sleep hygiene, or anxiety, depression, or just grounding. Yoga comes into that as well, breathing exercises – we have a weekly trauma-informed yoga group that many people find so helpful.

M: We're actually going to start running a psychoed group here first thing in the morning, before the gardening, as a trial. The first one is going to be for Farsi speakers, and we'll have an interpreter.

Linda: When I spoke with B, she told me how hard it was for her and the others during the Covid lockdowns, but that the support you were still able to provide was so vital.

M: During Covid lockdowns, people really suffered. We couldn't open the project, so instead we established an online meetup and online yoga classes as well. It was fantastic and really kept the group going. Some people couldn't do the technology, so we kept telephone contact. We'd had a little bit of money from the Maudsley charity, so we were able to send out a little kit to grow herbs on their windowsills. We even had some World Refugee Day celebrations, and someone came with a musician.

D: The pandemic was dreadful for people, absolutely awful. Their PTSD got worse because they were so isolated. We tried to think what could we do during this time to keep the group unified? What worked the best was yoga. So we decided that we could go online with yoga. It was quite a kerfuffle, just trying to get these people onto Zoom. Some of them were not at all techie; it was quite a challenge. With one member, it took weeks. I had to get someone to go around to his house and help him in the end.

We tried to really keep the idea of the group alive in everyone's minds. And we talked a lot with people about the importance of nature and the natural world, using that during the pandemic to settle themselves, to identify the natural spaces near to where they lived, and just talk to them about how it would help them to go out into those spaces. We've got various grounding techniques that we encourage about getting yourself into the present moment in nature. As Myriam said, we also sent them all some seeds and a planter and some soil. And that was hugely successful. People were sending photographs of what they were growing in their kitchens.

Linda: What do you feel you've learned from your work at the Grounding Project?

R: People gain so much from just the contact with other people. It's attachment isn't it? The importance and value of being able to offer something for individuals who can't necessarily engage in quite intensive talking therapies. This approach, through gardening, is so accessible, it's rewarding, it's natural. There are so many benefits to it. It's a treatment where you can engage regardless of how poorly you actually feel. You don't have to talk if you don't want to and the community understands because they share that with you. Even with the language barriers, people are still laughing and sharing things. They are not isolated and despairing.

D: I came here on a placement, as part of an MA in Refugee Care from Essex University in 2018 or 2019. Can't quite remember. I was really keen because I was just mad about nature and the natural world and was just getting very excited about horticultural therapy. This seemed like an ideal opportunity for me. I got very attached quite quickly. Normally, people hang around a bit after their placements, but the pandemic intervened in 2020. At that point, it was just me and Myriam – Myriam was the SLaM person, and I was the volunteer. And we realised that we would need to work really hard together to keep the group alive. I think that really consolidated my huge attachment to the group, wanting it to rise again after the pandemic. This is very, very special.

The energy in that organisation, Roots and Shoots, is also just phenomenal – we really do thrive on that, because they're constantly developing and changing and doing new projects. They really support us, they enable us to do the work that we do, because they're looking after our garden during the week while we're not here. They notice what's happening, water the plants and so on.

M: For me, I just feel it really puts me in touch with a sense of vitality that has developed within the group, with the garden and among group members. Connecting to nature and to other people is so fundamental to our being, and we are so lucky to benefit from this beautiful garden at Roots and Shoots and its incredibly supportive team here. It is a privilege and hugely rewarding to be involved in this project.

D: And members have been really interested in taking part in these interviews for the book too. The fact that it's not about their trauma and that they might help other people – it's actually very positive.[4]

***

We have heard from client members of the Grounding Project who have escaped persecution, violence and war in their homelands. But there are many millions of people worldwide who are displaced by climate-induced environmental disasters and ecocide. It is estimated that, over the next 25 years, up to 1 billion people worldwide will be driven from their homes to become climate migrants. As noted by Holmes (Chapter 1), this is currently affecting the Global South disproportionately, but these extreme events are increasingly occurring in Europe, North America, Australia ... As land is submerged, destroyed by wildfires, or becomes uninhabitable due to heat, the world's population will have to exist in ever-diminishing habitable space. Agriculture likewise, and we may struggle to produce enough food for everybody. Competition for land and resources will create tensions that, given the probable stress levels, will escalate and "conflicts over land and territory will likely proliferate as the accelerating climate crisis collides with rising geopolitical tensions" (Amrith, 2025). Massive loss of biodiversity is inevitable unless governments, corporations and big businesses commit to action. Many won't unless there is overwhelming public pressure. We cannot be complacent.

As borders in the Global North close to migrants and refugees from elsewhere, we privileged citizens who have contributed most to the ecological crisis seem adept at both denial and the narcissistic belief that "it won't happen to us." But it *is* happening. Wildfires and flooding, crop failures due to heat and aridity are here now. There is a further danger, too, that is facing Great Britain: coastal erosion caused by rising sea levels and impact of increasingly ferocious and frequent storms on tidal activity. It is not only Bangladesh, the Maldives and South Pacific island nations that are feeling it; with 15.5 per cent of the British coastline vulnerable, including several major cities. It is estimated that, by 2050, 1 million properties will be at risk of coastal erosion or flooding, with 1,600 km of major roads, 92 railway stations and other infrastructural features within 75 years (Rowe, 2022). Cardiff, the Welsh capital city, which currently has a population of around 495,000, is vulnerable and could lose two-thirds of its area to rising sea levels by 2050. The Committee for Climate Change research suggests that:

> The current record pace of human-induced climate change will mean that the UK's weather and climate will continue to change faster than ever over the decades ahead ... These changes will create risks across society that will reach unacceptable levels and therefore require proactive adaptation now. These risks include:

- *Increasing threats to nature and challenges to food production*
- *Extreme weather will more frequently disrupt the functioning of key infrastructure* [including water supply, roads and railways]
- *An increasing number of properties will be at risk of flooding or overheating* ... Towns and cities will become increasingly hot, with a large fraction of current buildings at risk of reaching uncomfortable and potentially dangerous temperatures in summer heatwaves
- *Future heat will risk the health and lives of vulnerable people* [an additional factor is] an increasingly vulnerable, ageing population. In heatwaves, many of our existing hospitals already struggle to operate effectively
- *Climate change will create challenges to economic prosperity*

(Committee for Climate Change Report, April 2025)

These changes are already happening, have been happening for decades. Vince (2021) argues that the only rational response is planned mass evacuations from the most vulnerable locations to newly constructed migrant cities in such places as the far north and Arctic. This hypothetical scenario is explored in the Danish television mini-series, *Families Like Ours* (2024, *Familier som vores*, written and directed by Thomas Vinterberg), where sea levels are rising at an alarming rate, and the government takes the desperate decision to evacuate the entire country, ripping apart families and communities as tensions escalate. These are not "migrants" looking for a better life, not asylum seekers escaping from war-torn countries or persecution in Africa or the Middle East, but Scandinavians known for rating highly in the index of happiness and comfort who are forced into exile as their homeland sinks under the seas, now homeless, losing their culture, their identity.

What will be the impact on our sense of identity as we lose forever many of the landscapes that shaped us, lost to coastal erosion, flooding, fire, or buried under densely constructed housing? Our homes, gardens, allotments? Our communities? With the far-right stoking hatred toward asylum seekers, what reception will be given to a great number of "incomers" who are internally displaced British climate refugees? I would like to dream that we become more sensitive to and compassionate towards migrants and people seeking refuge who are so frequently demonised. Yet I fear that, as we scrabble for ever-diminishing resources and attempt to protect what we perceive as "ours," we might resort to reactive trauma responses, further splitting and projection, escalating sibling rivalry prompted by limited resources and the "narcissism of minor differences" (Freud, 1930).

It is a nightmare scenario, and one that can only be mitigated by pulling together now, recognising that we are all in the same small boat. We can all play our part. At the Grounding Project, people find community in nature. Now we need community to save nature, the wonderful world we have laid claim to with all its magnificent biodiversity. John Bowlby, and Bowlby Centre founder John Southgate, would surely put their names to that.

Notes

1 All clients of the Grounding Project who took part in these interviews gave informed consent for their interviews to be recorded and transcribed. They were free to decline taking part with no consequences, and to choose to not answer any questions they felt uncomfortable with. They then approved those sections included here after which both recordings and transcripts were deleted.
2 Linden and Grut (2022).
3 Herman (1992).
4 My huge appreciation to everyone at the Grounding Project for welcoming me in and being part of this project. Myriam, Rebecca and Dawn did a great deal to make this chapter possible, from liaising with SLaM, devising an agreement for clients taking part, discussing the project with client members, sitting in on interviews to ensure clients felt safe and to raise additional questions, and much more.

References

Achotegui, J. (2019). Migrants living in very hard situations: Extreme migratory mourning (the Ulysses syndrome). *Psychoanalytic Dialogues. 3*: 252–268.

Amrith, S. (2025). Your land is our land: Property, Power, and the coming age of territorial conflict. *Foreign Affairs*, July/August 2025. Available at: www.foreignaffairs.com/reviews/your-land-our-land-amrith?s=EDZZZ005ZX&utm_medium=newsletters&utm_source=fatoday&utm_campaign=Trump%20Is%20Breaking%20American%20Intelligence&utm_content=20250702&utm_term=EDZZZ005ZX#

Blake, W. (1804). *Milton: A Poem*. London: Shambala, 1978.

Climate Change Committee (2025). Progress in adapting to climate change: Report to parliament. *The CCC*, 30 April 2025. Available at: www.theccc.org.uk/publication/progress-in-adapting-to-climate-change-2025/ (accessed 18 August 2025).

Families Like Ours (2024). StudioCanal/BBC, March 2025. Television mini-series.

Freud, S. (1930). *Civilisation and Its Discontents. S.E., 21*. London: Hogarth.

Herman, J.L. (1992). *Trauma and Recovery: From Domestic Abuse to Political Terror*. New York: Basic Books.

Linden, S., & Grut, J. (2022). *The Healing Fields: Working with Psychotherapy and Nature to Rebuild Shattered Lives*. London: Frances Lincoln, *in association with The Medical Foundation for the Care of Victims of Torture*.

Roberts, A. (2017). *Tamed: Ten Species that Changed Our World*. London: Penguin.

Rowe, M. (2022). The rising threat of UK coastal erosion. *Geographical*, 14 September 2022. Available at: https://geographical.co.uk/climate-change/the-rising-threat-of-uk-coastal-erosion (accessed 18 August 2025).

Rust, M.-J. (2020). *Towards an Ecotherapy*. London: Confer Books.

Snow, L. (2019). *War Gardens: A Journey Through Conflict in Search of Calm*. London: Quercus.

Stuart-Smith, S. (2020). *The Well Gardened Mind: Rediscovering Nature in the Modern World*. London: William Collins.

Táíwò, O.O. (2025). *Reconsidering Reparations: Why Climate Justice and Constructive Politics are Needed in the Wake of Slavery and Colonialism*. Chicago: Haymarket Books.

Vince, G. (2021). *Nomad Century: How to Survive the Climate Upheaval*. London: Allen Lane.

Williamson, L. (2025) BBC witnesses Israeli settlers' attack on Palestinian farm in West Bank. *BBC News*, 18 August 2025. Available at: www.bbc.co.uk/news/articles/cewy88jle0eo (accessed 18 August 2025).

Postscript

Linda Cundy

Attachment

This book, based on the 27th John Bowlby Memorial Conference, has explored the major themes of Attachment Theory. Bowlby reminded us that we are animals, social mammals with a repertoire of instinctual behaviours, who turn to others in infancy and throughout life for comfort, reassurance and protection from danger. This instinct promotes individual survival and has ensured the survival of our species. During the course of maturation, most of us develop the capacity to then nurture and protect others, to provide a secure base for them. Our relationships matter to us, and we grieve when we lose someone or something close. However, our inner resources to feel compassion and take action to help others may be limited or conflictual.

Resources and rivalry

These are troubled times. There have always been wars, often stoked by historical conflicts, hatred and desire for revenge, but also, overtly or covertly, relating to envy and a primitive desire to take what the other possesses – access to "mother," access to mineral rights and other natural resources. The presence of significant natural gas reserves in the Ukraine region of the Black Sea may be one motivation for Russia's invasion in 2022. A licence was awarded by Israel to British Petroleum, among other companies, to explore and extract untapped gas reserves in the eastern Mediterranean off the coast of Gaza. BP pulled out – not because of climate concerns about continuing use of fossil fuels, but because the situation is too risky given the conflict in that region. Greenland has significant unmined deposits of minerals and, possibly, untapped reserves of gas and oil that no doubt make it irresistibly attractive to powerful neighbours intent on acquiring these, despite the feelings of its native inhabitants. Tibet also has vast natural and mineral resources, including lithium, used to produce green energy, and the major rivers formed by water running off the Himalayas that are harnessed by the Chinese government to generate hydroelectricity. These natural treasures could help reduce reliance on fossil fuels, but

DOI: 10.4324/9781003733942-7

evidence suggests the Chinese government's approach takes little consideration of ecosystems, let alone the rights and well-being of the Tibetan and other minority peoples. Indeed, "the potential for Chinese weaponization of clean energy, given Beijing's production dominance in minerals as well as in solar panels and batteries, is a key element of this new age of intense geopolitical rivalry" (Bordoff & O'Sullivan, 2025).

In families, sibling rivalry is a product of insecure attachment occurring when children sense there are not enough resources – love, affection, protection – to go around. What we are seeing are dangerous examples of global sibling rivalry between factions and nations; "it's not fair – he's got more than me," "I want what they've got" – and with supplies of these resources dwindling it is likely that conflict will proliferate as determination escalates to grab them at any cost. Wars do great evil, including fuelling the eco-crisis and undoing much of the good that has been achieved in tackling global warming. We have inter-generational trauma. War begets war.

Meanwhile, we increasingly see splitting, Us and Them, a shift towards nationalism in politics and demonising of minority groups. Psychoanalytic psychotherapist and founder of the Climate Psychology Alliance Paul Hoggett refers to the rise in reactionary states of mind as the "politics of the armed lifeboat," an image coined by Parenti (2011): fearing competition for our own scarce resources, "we" survive by closing our borders and our hearts to others, leaving them out in the cold – or to drown. He writes:

> I believe that a lurch into authoritarian nationalism is a much more likely initial response to the deepening climate crisis in the coming decades. This is what we should be frightened of, not catastrophist fantasies of panicked crowds fighting in the aisles of supermarkets as the shelves run dry. The first signs of this imminent authoritarianism could be discerned in the rise of populist and nationalist movements.
>
> (Hoggett, 2023, p. 98)

Also in the erosion of the right to free reporting, and the right to peaceful protest …

We lose *our* humanity by dehumanising others. Nobody behaves their best when under extreme duress. We know from working with our traumatised patients that states of high agitation, fear, rage and hatred need to be down-regulated in order to help them access their capacity to mentalise, to reflect, to feel compassion, to make informed judgements. But turning down the world's emotional heat while global temperatures reach ever-higher is a big ask.

A reminder

Somewhere in our history, our species laid claim to all the planet's resources and, as such, we have taken on stewardship of the Earth. Our technologies and colonial, consumerist cultures have created the environment we now inhabit – depleted, polluted, endangered – but still magnificent and precious. Now it is our responsibility to save what can be saved, to protect what must be protected, to respect and value what survives for future generations. More than caretakers of Earth, we need to be care-givers, treasuring and protecting our planet against further violation. No longer greedy, demanding children always wanting more, with no realistic sense of consequences; we must make that developmental leap to become creatures who give to, not take from, our environment, and who share with our neighbours, both human and other-than-human.

It will not be easy. Initially, there must be a huge financial investment in new technologies, and we must radically re-think our priorities. There is a reluctance to make the necessary changes in the form of apathy from many, but a more aggressive resistance from those with huge stakes in fossil fuels. There have been many permanent losses, with more to come. Meanwhile, so many of us feel helpless to make a difference.

We are perhaps the most adaptable creatures on this planet; we have evolved physically and socially to inhabit almost anywhere: rainforests, deserts, small coral islands, green pasture land, frozen tundra, high mountains, swamps, and concrete jungles. We adapt quickly, too, and in some respects that is part of the problem. We seem to acclimatise to increasingly hot, dry or wet weather, and, unless our homes and communities are devastated, we clean up after the floods and cyclones, draw the curtains against the rising temperatures and get on with things (as in the short story by Kevin McCabe in Maggie Turp's Chapter 4 in this volume). Yet it is impossible to dissociate; our nervous systems are registering the build-up of anxiety and dangers even if our thoughts are elsewhere.

Concern for the environment is not new. In 1962, Rachel Carson published *Silent Spring* (Carson, 1962), highlighting the major damage caused by the pesticide DDT. Eight years later, Canadian singer Joni Mitchell released "Big Yellow Taxi," one of her most-loved songs about the same subject (Mitchell, 1970). And musician Richie Havens sang "What Are You Going to Do About Me" (Havens, 1971) one of the great protest songs about pollution, deforestation and injustice. If you are not familiar with it, I encourage you to watch him perform this song on YouTube. His powerful message is a call to action as relevant now as when it was first recorded.

We did care, deeply, about these issues at that time.

In 1972, more than 50 years ago, the great American psychoanalyst and specialist in schizophrenia Harold Searles wrote a powerful paper, "Unconscious Processes in Relation to the Environmental Crisis." He cites pollutants in the oceans and atmosphere, radioactive waste and unsustainable overpopulation. He writes,

> our fear, envy, and hatred of formidable Oedipal rivals makes us view with large-scale apathy their becoming polluted into extinction … [However, our] envy of the more favored rival is provided vicarious satisfaction by the simple leveling effect of the universal environmental pollution.
>
> (Searles, 1972, pp. 364–365)

The following year saw the publication of the hugely influential book, *Small is Beautiful*, (Schumacher, 1973) about the impact of globalisation, highlighting the inescapable fact that natural resources are finite, and urging the development of sustainable technologies.

There were some successes: DDT was eventually banned in most countries. But then many of us took our eye off the ball. I confess that I was one of them. But now we can no longer deny the impact of climate change, species losses and ecological degradation – we may disavow, but the reality is all around us; we breathe it in.

Now

In the Anthropocene, we are just one species sharing the world's resources, but we exploit far more than any other creature. We also have the greatest responsibility to make amends for our exploitation, the damage we have caused. We all have to change how we live, focus on what we most value – short-term consumerism and financial profit – or relationships, secure societies and the health of the planet with all its marvellous diversity of inhabitants.

Searles again:

> I postulate that an ecologically healthy relatedness to our non-human environment is essential to the development and maintenance of our sense of being human, and that such a relatedness has become so undermined, disrupted and distorted, concomitant with the ecological deterioration, that it is inordinately difficult for us to integrate the feeling experiences, including the losses inescapable to any full-fledged human being.
>
> (1972, p. 367)

Like insecure toddlers with a huge box of toys but unwilling to share them, and envious of what others have, we have been the greedy grabbers of fossil fuels, minerals, land and ocean resources. Rather than dehumanising our fellow *Homo sapiens*, perhaps we should be reframing our perceptions of all the Earth's inhabitants and gifts.

In 2017, the Whanganui River in New Zealand was granted the same legal rights as a human being under New Zealand law. The Indigenous Maori people respect the river as an ancestor. This legal status means that anybody causing damage to the Whanganui can be prosecuted. More recently, in 2025, the Taranaki Mounga mountain was likewise granted legal personhood for the same reason.

In India, the holy river Ganges and its Yamuna tributary are recognised as living entities and have also been granted personhood status in India, entitling them to the same rights of protection under law as humans.

In 2010, a World People's Conference on Climate Change and the Rights of Mother Earth held in Bolivia agreed a Universal Declaration of the Rights of Mother Earth (2010), including Obligations of human beings to Mother Earth. It makes for an inspiring if idealistic read. Bolivia also passed the world's first "law of Mother Earth" to protect the rights of its natural environment. Sadly, since that time, there have been many violations: "Since the passing of these laws, the Bolivian government has pushed for numerous projects which involve the invasion of protected areas and the violation of Indigenous peoples' rights" (Muñoz, 2023) claims Muñoz, pointing to the endemic poverty contrasting with rich mineral and other resources, enabling governments and "corporate-driven neoliberal" interests to undermine the legislation. Let us hope that New Zealand and India do not succumb to similar pressures.

South African environmental lawyer and author of *Wild Law* (Cullinan, 2003), Cormac Cullinan was awarded the Shackleton Medal for Polar Research. In an interview with journalist Donna Ferguson he urges the adoption of powerful and effective laws to protect the Antarctic continent, pointing out that it has no Indigenous population and, thus, no government to protect it from the impact of mineral exploitation and climate change. He proposes that Antarctica be internationally recognised as having "legal personhood" as "a being in its own right" that would enable it to take legal action against those who endanger it (Ferguson, 2025). I am reminded of the Bowlby Centre's founder John Southgate and his involvement with the Council for Posterity (see Preface).

However, "In February 2024, a British government official told the United Nations (UN) that the United Kingdom (UK) would never give rights to nature." During a debate on preliminary negotiations for the UN Environment Assembly in Nairobi, the British delegate from the Department for Environment, Food, and Rural Affairs (Defra) said,

> The UK's firm position is that rights can only be held by legal entities with a legal personality. We do not accept that rights can be applied to nature or Mother Earth. While we recognise that others do, it is a fundamental principle for the UK and one from which we cannot deviate.
>
> (Dent, 2024)

Crisis and opportunity

The planet faces an almighty challenge. We can lower our eyes and leave it to others to clear up the mess, or rise to the challenge and view it as an opportunity to work towards a better world. We can be silenced by vested interests, cling to our old consumerist addictions, succumb to despair, turn against our fellow humans, or play a small part in creating a more equitable world. As consumers, we have power. We can shrug or step up, pile the pressure on governments and corporations to do what must be done. We can oppose the prevailing narrative that demonises and divides.

Economist Chris Goodall's valuable book *What We Need to Do Now* (Goodall, 2020) highlights action needed at a policy level to reach zero-carbon emission targets, but also highlights the positive steps each individual can take to play their part. From better home insulation to choosing green energy, buying less fast fashion and updating electronic equipment less frequently, cutting down on eating meat and avoiding air-freighted foods where possible, to voting with the environment in mind and putting pressure on policy-makers at all levels – as Greta Thunberg said, *no one is too small to make a difference*. There have been so many losses, and there will be many more – it's too late to turn things around. But our actions now can still make a difference.

Despite my perception that public engagement with environmental issues waned at the end of the 20th century and the start of this one, there have been significant positive developments. To some extent, the concern has moved from the fringes of a protest movement to the mainstream, to the laboratories investigating new, more sustainable materials, to the boardrooms of big businesses developing green technologies, and to the wider discourse. As Solnit reminds us,

> in the past two decades, climate actions and energy transitions have shifted our [global warming] trajectory downward dramatically. Had there been no climate action, we might be hurtling toward 4 degrees or more of warming, far deeper in catastrophe than we are now, and far less equipped to limit it.
> (Solnit, 2023, p. 7)

Do you remember the time before recycling? When the go-to for packaging was plastic, and every purchase came with a free plastic bag? When everyone, even in cities, drove their children to school or to the local supermarket? Before electric vehicles? Before eco settings on household devices and environmentally friendly products were made sustainably? There have been major advances in recent decades without which the Earth would be in far greater danger; yet it is nothing like enough. Climate change is exponential – but so are the new technologies to limit it. We now have direct carbon capture, off-setting, hydrogen batteries, plastic-eating enzyme technology that can dramatically reduce plastic pollution … We have options now that were not available to us just a few years ago. But we can't be complacent or leave it to scientists, businesses and governments to find and implement solutions to the life-endangering problems we all have had a hand in creating. We must take our share of responsibility.

Those of us with the greatest privileges have the greatest responsibility, but we also have more choices available to us. If possible, we can choose to buy organic foods – or even grow our own. Omnivores can opt to eat less meat or become vegetarians; vegetarians can commit to eating less dairy, more vegan. We may be able to afford ethically sourced and environmentally friendly goods, drive less, fly less, offset the carbon emissions we generate. However, for many people, even in the Global North, these are extravagances beyond their wildest dreams. But anyone can volunteer in projects to improve their local environment, anyone can be mindful about recycling, use of water and energy. And anyone at all could have ideas that make a difference; scientists, philosophers and footballers, domestic and construction workers, hospital porters and cab drivers, children, Indigenous peoples, neurodivergent individuals, artists, people seeking refuge, and even psychotherapists. Government, schools, universities, community projects can all encourage thinking outside the box.

***

I recently attended a social gathering, sitting around a big garden table with people of all ages on a beautiful early summer day. From the initial light chat about the weather, we began to drift into a more serious conversation about global warming until a teenage girl spoke out: "We are destroying everything. It would be best if humans were all wiped out and the world could carry on without us." An awkward hush around the table was followed by humorous protestation and a few rather patronising remarks intended to restore the equilibrium and sunny mood.

I recognised in her myself at her age, listening to Richie Havens and Joni Mitchell, joining CND and going to protest rallies, raging about humanity's cruelty to other people and to other creatures. I spoke with her privately later and told her that I understood her anger and hopelessness. Perhaps we will eventually destroy ourselves, but if that's the case we will unfortunately be taking most of the planet's riotous biodiversity with us. My advice was, rather than despair, and even if our efforts are ultimately futile, get involved and *do* something because that's the right thing to do. With her sensitivity and passion, this young woman has so much to contribute to the protection of our non-human neighbours on the planet. I had a hunch that belonging to a group of activists would give her a much-needed sense of belonging and self-respect, too – it could become a secure base for her as she navigates adolescence and her early 20s. But perhaps I am over-identified, projecting my own adolescent needs onto her!

You don't have to join Extinction Rebellion if that is not your thing (but if it is, please do). As I told my young relative, the antidote to helplessness is action, especially as part of a group or movement. After exploring the context of their distress and anger I would say the same to patients – *do something!*

Ideals and values are an important part of identity. I realise as I write this that I rarely ask clients about theirs.

***

Psychotherapy

The aim of an attachment-informed psychotherapy is to help our clients develop secure attachment and to thrive, rather than merely survive. This feels ironic when the environment we inhabit is fraught with dangers. However, as Jacobs points out (Chapter 5), people who are securely attached have a greater capacity to manage their anxiety, to mentalise, to keep thinking in stressful situations. They are also inclined to bring people together, to feel empowered, and to motivate. These qualities are needed now and in the future to confront the enormity of the task facing humanity. Therapy helps people connect with and nurture their internal resources, gain perspective and make informed choices. Along the way it challenges defensive narcissism – the difficulty recognising that other people are individuals in their own right with their own worries, needs, skills, and that one's behaviour has an impact on others. It enables patients to be more present, in their bodies and in the world. Individual therapy alone will not save the planet, but it is a contribution we psychotherapists can make to creating healthier, more respectful relationships between people, and between people and the world we inhabit. I do not have the answer to how we build a world that is fairer, more caring, with increased motivation to protect the environment, where there is the will to create greater secure attachment at a societal level. But there are people out there who are thinking about this, trying to bring about a profound social change.

> We have all the tools to build our societies anew, reform our ways of thinking, fix the inequalities and end the discriminations, and choose earnest wisdom over snippets of information, choose empathy over hatred, choose humanism over tribalism, yet we don't have much time or room for error while we are losing our planet, our only home.
>
> (Shafak, 2020, p. 90)

Climate psychology is rapidly developing and urgently needed with therapeutic practitioners from many disciplines uniting to explore the impact of global warming on mental health and relationships, to understand and challenge the defences so often evident in society, and to reflect on ethical psychotherapy practice (see, for instance, Anderson et al., 2024; Rust, 2020). It is imperative that attachment-informed psychotherapists join the discussion, address the reality of climate change and play our part in working towards the creation of more secure societies.

When I organised this conference for the Bowlby Centre, my hope was to bring people together, to recognise the unique place of Attachment Theory in addressing the climate and environmental crisis and to share ideas and energy, to motivate. I wanted us to think together about the impact of eco-anxiety, not only on our clients but on ourselves, and to acknowledge the challenges of being useful psychotherapists without imposing our own fears. I hoped it would be a call to action. I hold the same hope for this book. If psychotherapy can facilitate "earned security" for individuals, it must have a role in working towards more secure and just societies, a healthier planet. For myself, I'd rather aim for Bowlby's vision of social justice through psychologically informed social policies, and Jeremy's, of an attachment-informed politics and environmentally committed policies – than the politics of the armed lifeboat, even though that will require sacrifices.

I end with these words from Amir Nizar Zuabi, creative director of "The Herds," a massive international travelling puppet project aiming to raise awareness of the environmental emergency and to inspire action:

> We live in a world with constant alarm bells and warnings yet the urgency is not registering. We want to disturb this indifference ... It needs to happen to *us*, not someone else far away. It needs to happen somewhere we feel safe so we understand that we are *not* safe ... I know we are just water dripping on a stone but it's crucial to continue dripping because slowly, over time, water shapes the stone.
>
> (Zuabi, 2023, TED Talk)

References

Anderson, J., Staunton, T., O'Gorman, J., & Hickman, C. (Eds.) (2024). *Being a Therapist in a Time of Climate Breakdown*. London: Routledge.

Bordoff, J., & O'Sullivan, M.L. (2025). The return of the energy weapon: An old tool creating new dangers. *Foreign Affairs*, 21 October 2025. Available at: http://www.foreignaffairs.com/united-states/return-energy-weapon-bordoff-osullivan?utm_medium=newsletters&utm_source=fatoday&utm_campaign=The%20Return%20of%20the%20Energy%20Weapon&utm_content=20251027&utm_term=L#

Carson, R. (1962). *Silent Spring*. Boston, MA: Houghton Mifflin.

Cullinan, C. (2003). *Wild Law: A Manifesto for Earth Justice*. London: Green Books.

Dent, C. (2024). Rights of nature, or wrongs to nature? The denial of legal personhood to nature in the UK. *Earth Law Centre.* Available at: www.earthlawcenter.org/blog-entries/2024/4/rights-of-nature-or-wrongs-to-nature-the-denial-of-legal-personhood-to-nature-in-the-uk#:~:text=Rights%20of%20Nature%20or%20Wrongs,to%20Nature%20in%20the%20UK&text=In%20February%202024%2C%20a%20British,never%20give%20rights%20to%20Nature

Ferguson, D. (2025). Antarctica "too wild for humans to rule," says Shackleton medal winner. *The Guardian*, 7 June 2025. Available at: www.theguardian.com/environment/2025/jun/07/cormac-cullinan-shackleton-medal-antarctica-environmental-continent-legal-status?CMP=share_btn_url

Goodall, C. (2020). *What We Need to Do Now for a Zero Carbon Future*. London: Profile.

Havens, R. (1971). What you going to do about me. In *The Great Blind Degree [Album]*. Culver City, CA: Stormy Forest Records. Available at: www.youtube.com/watch?v=1utb4rvuk6U

Hoggett, P. (2023). *Paradise Lost? The Climate Crisis and the Human Condition*. Winterhuge: Simplicity Institute.

Mitchell, Joni (1970). Big Yellow Taxi. In *Ladies of the Canyon [Album]*. Los Angeles, CA: Reprise Records.

Muñoz, Lorna (2023). Bolivia's Mother Earth laws: Is the ecocentric legislation misleading? *ReVista: Harvard Review of Latin America*, 6 February 2023. Available at: https://revista.drclas.harvard.edu/bolivias-mother-earth-laws-is-the-ecocentric-legislation-misleading/ (accessed 8 November 2025).

Parenti, C. (2011). *Tropic of Chaos: Climate Change and the New Geography of Violence*. New York: Bold Type Books.

Rust, M.-J. (2020). *Towards an Ecopsychotherapy*. London: Confer Books.

Schumacher, E.F. (1973). *Small is Beautiful: A Study of Economics as If People Mattered*. London: Vintage.

Searles, H. (1972). Unconscious processes in relation to the environmental crisis. *Psychoanal. Rev.* 59(3): 361–374.
Shafak, E. (2020). *How to Stay Sane in an Age of Division*. London: Profile.
Solnit, R. (2023). Difficult is not the same as impossible. In Solnit, R., & Lutunatabua, T.Y. (Eds.), *Not Too Late: Changing the Climate Story from Despair to Possibility* (pp. 3–10). Chicago, Illinois: Haymarket.
Thunberg, G. (2019). *No One is Too Small to Make a Difference*. London: Penguin.
Universal Declaration of the Rights of Mother Earth (2010). Available at: www.garn.org/wp-content/uploads/2024/02/ENG-Universal-Declaration-of-the-Rights-of-Mother-Earth.pdf
Zuabi, Amir Nizar (July 2023). The Herds, a vast act of theater to spark climate action. *TED Talk*. Available at: www.ted.com/talks/amir_nizar_zuabi_the_herds_a_vast_act_of_theater_to_spark_climate_action

Index

Note: *italic* page numbers indicate figures; page numbers followed by n refer to notes.

For Product Safety Concerns and Information please contact our EU
representative GPSR@taylorandfrancis.com
Taylor & Francis Verlag GmbH, Kaufingerstraße 24, 80331 München, Germany

www.ingramcontent.com/pod-product-compliance
Ingram Content Group UK Ltd.
Pitfield, Milton Keynes, MK11 3LW, UK
UKHW020001200926
475702UK00015B/2408

9781041227847